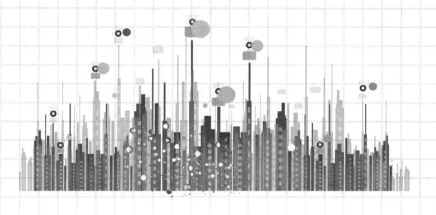

Python 数据可视化

方法、实践与应用

王振丽 ◎ 编著

清华大学出版社
北京

内 容 简 介

本书循序渐进、深入讲解了使用 Python 语言实现数据可视化分析的核心知识，并通过具体实例的实现过程演练了数据可视化分析的方法和流程。全书共 10 章，内容包括数据采集、使用数据库保存数据、绘制散点图和折线图、绘制柱状图、绘制饼状图、绘制其他图形以及商业应用——电影票房数据可视化、房地产市场数据可视化、交通数据可视化、招聘信息可视化。讲解简洁而不失深度，内容丰富、全面，历史资料翔实完整。本书以极简的文字介绍了复杂的案例，易于理解。本书适用于已经了解了 Python 语言基础语法的读者，也适用于希望进一步提高自己 Python 开发水平的读者，还可以作为大专院校相关专业的师生用书和培训机构的教材。

本书封面贴有清华大学出版社防伪标签，无标签者不得销售。
版权所有，侵权必究。举报：010-62782989，beiqinquan@tup.tsinghua.edu.cn。

图书在版编目(CIP)数据

Python 数据可视化方法、实践与应用/王振丽编著. —北京：清华大学出版社，2020.12(2022.2 重印)
ISBN 978-7-302-56867-4

Ⅰ. ①P… Ⅱ. ①王… Ⅲ. ①软件工具—程序设计 Ⅳ. ①TP311.561

中国版本图书馆 CIP 数据核字(2020)第 226275 号

责任编辑：魏　莹
装帧设计：杨玉兰
责任校对：周剑云
责任印制：丛怀宇

出版发行：清华大学出版社
网　　址：http://www.tup.com.cn, http://www.wqbook.com
地　　址：北京清华大学学研大厦 A 座　　邮　　编：100084
社 总 机：010-62770175　　邮　　购：010-62786544
投稿与读者服务：010-62776969, c-service@tup.tsinghua.edu.cn
质量反馈：010-62772015, zhiliang@tup.tsinghua.edu.cn

印 装 者：大厂回族自治县彩虹印刷有限公司
经　　销：全国新华书店
开　　本：185mm×260mm　　印　张：20.5　　字　数：498 千字
版　　次：2020 年 12 月第 1 版　　印　次：2022 年 2 月第 2 次印刷
定　　价：79.00 元

产品编号：087732-01

前言

互联网的飞速发展伴随着海量信息的产生，而海量信息的背后对应的则是海量数据。如何从这些海量数据中获取有价值的信息来供人们学习和工作使用，这就不得不用到大数据挖掘和分析技术。数据可视化分析作为大数据技术的核心一环，其重要性不言而喻。

随着云时代的来临，数据可视化分析技术将具有越来越重要的战略意义。大数据已经渗透到每一个行业和业务职能领域，逐渐成为重要的生产要素，人们对于海量数据的运用将预示着新一轮生产率增长和消费者盈余浪潮的到来。数据可视化分析技术将帮助企业用户在合理时间内攫取、管理、处理、整理海量数据，为企业经营决策提供积极的帮助。数据可视化分析作为数据存储和挖掘分析的前沿技术，已广泛应用于物联网、云计算、移动互联网等战略性新兴产业。虽然数据可视化分析目前在国内还处于初级阶段，但是其商业价值已经显现出来，特别是有实践经验的数据可视化分析人才更是成为各企业争夺的热门。为了满足日益增长的数据可视化分析人才需求，很多大学开始尝试开设不同程度的数据可视化分析课程。"数据可视化分析"作为大数据时代的核心技术，必将成为高校数学与统计学专业的重要课程之一。

本书的特色

1．内容全面

本书详细讲解 Python 数据可视化分析所需要的开发技术，循序渐进的讲解了这些技术的使用方法和技巧，帮助读者快速步入 Python 数据分析的高手之列。

2．实例驱动教学

本书采用理论加实例的教学方式，通过对这些实例实现了对知识点的横向切入和纵向比较，让读者有更多的实践演练机会，并且可以从不同的方位展现一个知识点的用法，真正实现了拔高的教学效果。

3．详细介绍了数据分析的流程

本书从一开始便对数据分析的流程进行了详细介绍，而且在讲解中结合了多个实用性很强的数据分析项目案例，带领读者掌握 Python 数据分析的相关知识，以解决实际工作中的数据分析问题。

4．二维码布局全书，扫码后可以观看讲解视频

本书正文的每一个二级目录都有一个二维码，通过二维码扫描可以观看讲解视频，既包括实例讲解也包括教程讲解，对读者的开发水平实现了拔高处理。

5. 贴心提示和注意事项提醒

本书根据需要在各章安排了很多"注意""说明"和"技巧"等小板块,让读者可以在学习过程中更轻松地理解相关知识点及概念,更快地掌握个别技术的应用技巧。

本书的内容

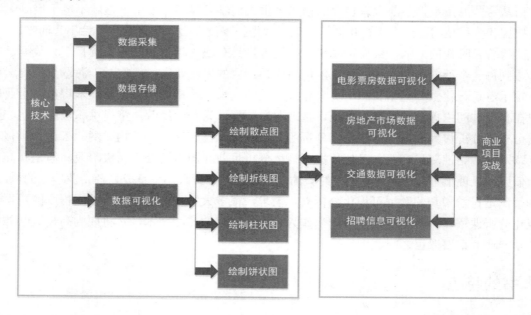

本书的读者对象

- 软件工程师。
- Python 语言初学者。
- Python 初学者和自学者。
- 专业数据分析人员。
- 数据库工程师和管理员。
- 研发工程师。
- 大学及中学教育工作者。

致谢

本书在编写过程中,得到了清华大学出版社各位专业编辑们的大力支持,正是各位专业人士们的求实、耐心和效率,才使得本书能够在这么短的时间内出版。另外,也十分感谢我的家人给予的巨大支持。本人水平毕竟有限,书中存在纰漏之处在所难免,诚请读者提出宝贵的意见或建议,以便修订并使之更臻完善。

编　者

目录

第 1 章 数据采集 1

1.1 处理网络数据 2
- 1.1.1 解析 HTML 和 XML 数据 2
- 1.1.2 处理 HTTP 数据 11
- 1.1.3 处理 URL 数据 17

1.2 网络爬虫技术 21
- 1.2.1 网络爬虫基础 21
- 1.2.2 使用 Beautiful Soup 爬取网络数据 22
- 1.2.3 使用 XPath 爬取网络数据 24
- 1.2.4 爬取体育新闻信息并保存到 XML 文件 26
- 1.2.5 爬取 XX 百科 29

1.3 使用专业爬虫库 Scrapy 32
- 1.3.1 Scrapy 框架基础 33
- 1.3.2 搭建 Scrapy 环境 34
- 1.3.3 创建第一个 Scrapy 项目 34
- 1.3.4 爬取某电影网的热门电影信息 38
- 1.3.5 爬取某网站中的照片并保存到本地 42
- 1.3.6 爬取某网站中的主播照片并保存到本地 43

第 2 章 使用数据库保存数据 45

2.1 操作 SQLite 3 数据库 46
- 2.1.1 sqlite3 模块介绍 46
- 2.1.2 使用 sqlite3 模块操作 SQLite 3 数据库 53
- 2.1.3 SQLite 和 Python 的类型 56

2.2 操作 MySQL 数据库 60
- 2.2.1 搭建 PyMySQL 环境 61
- 2.2.2 实现数据库连接 62
- 2.2.3 创建数据库表 62

2.3 使用 MariaDB 数据库 63
- 2.3.1 搭建 MariaDB 数据库环境 63
- 2.3.2 在 Python 程序中使用 MariaDB 数据库 66
- 2.3.3 使用 MariaDB 创建 MySQL 数据库 68

2.4 使用 MongoDB 数据库 71
- 2.4.1 搭建 MongoDB 环境 71
- 2.4.2 在 Python 程序中使用 MongoDB 数据库 72

2.5 使用 ORM(对象关系映射)操作数据库 75
- 2.5.1 Python 和 ORM 75
- 2.5.2 使用 SQLAlchemy 76
- 2.5.3 使用 mongoengine 80

第 3 章 绘制散点图和折线图 83

3.1 绘制散点图 84
- 3.1.1 绘制一个简单的点 84
- 3.1.2 添加标题和标签 84
- 3.1.3 绘制 10 个点 85
- 3.1.4 修改散点的大小 86
- 3.1.5 设置散点的颜色和透明度 87
- 3.1.6 修改散点的形状 87
- 3.1.7 绘制两组数据的散点图 88
- 3.1.8 为散点图设置图例 89
- 3.1.9 自定义散点图样式 89
- 3.1.10 使用 pygal 绘制散点图 91

3.2 绘制折线图 92
- 3.2.1 绘制最简单的折线 92
- 3.2.2 设置标签文字和线条粗细 93
- 3.2.3 绘制 1000 个点组成折线图 94
- 3.2.4 绘制渐变色的折线图 95
- 3.2.5 绘制多幅子图 96

3.2.6 绘制正弦函数和余弦函数曲线 97
3.2.7 绘制 3 条不同的折线 100
3.2.8 绘制浏览器市场占有率变化折线图 101
3.2.9 绘制 XY 线图 102
3.2.10 绘制水平样式的浏览器市场占有率变化折线图 103
3.2.11 绘制叠加折线图 104
3.2.12 绘制某网站用户访问量折线图 105
3.3 绘制其他类型的散点图和折线图 ... 106
3.3.1 绘制随机漫步图 106
3.3.2 大数据可视化分析某地的天气情况 110
3.3.3 在 Tkinter 中使用 Matplotlib 绘制图表 113
3.3.4 绘制包含点、曲线、注释和箭头的统计图 115
3.3.5 在两栋房子之间绘制箭头指示符 117
3.3.6 根据坐标绘制行走路线图 ...118
3.3.7 绘制方程式曲线图 120
3.3.8 绘制星空图 122
3.4 绘制 BTC(比特币)和 ETH(以太币)的价格走势图 122
3.4.1 抓取数据 122
3.4.2 绘制 BTC/美元价格曲线 ... 123
3.4.3 绘制 BTC 和 ETH 的历史价格曲线图 124
3.5 Flask+pygal+SQLite 实现数据分析 125
3.5.1 创建数据库 125
3.5.2 绘制统计图 126

第 4 章 绘制柱状图 129
4.1 绘制基本的柱状图 130
4.1.1 绘制只有一个柱子的柱状图 130

4.1.2 绘制有两个柱子的柱状图 130
4.1.3 设置柱状图的标签 132
4.1.4 设置柱状图的颜色 135
4.1.5 绘制堆叠柱状图 136
4.1.6 绘制并列柱状图 137
4.1.7 绘制 2002—2013 年网页浏览器使用变化柱状图 138
4.1.8 绘制直方图 139
4.1.9 绘制横向柱状图 140
4.1.10 绘制有图例横向柱状图 141
4.1.11 绘制分组柱状图 142
4.1.12 模拟电影票房柱状图 144
4.1.13 绘制正负柱状图 145
4.1.14 绘制不同商品销量的统计柱状图 145
4.2 可视化分析掷骰子游戏的结果次数 146
4.2.1 使用库 pygal 实现模拟掷骰子功能 147
4.2.2 同时掷两个骰子 148
4.3 可视化分析最受欢迎的开源项目 150
4.3.1 统计前 30 名最受欢迎的 Python 库 150
4.3.2 使用 pygal 实现数据可视化 152
4.4 可视化统计显示某网店各类口罩的销量 154
4.4.1 准备 CSV 文件 154
4.4.2 可视化 CSV 文件中的数据 155
4.5 数据挖掘:可视化处理文本情感分析数据 156
4.5.1 准备 CSV 文件 156
4.5.2 可视化两个剧本的情感分析数据 157

第 5 章 绘制饼状图 161
5.1 绘制基本的饼状图 162

5.1.1	绘制简易的饼状图................162	
5.1.2	修饰饼状图............................164	
5.1.3	突出显示某个饼状图的部分................................165	
5.1.4	为饼状图添加图例................166	
5.1.5	使用饼状图可视化展示某地区程序员的工龄....................166	
5.1.6	绘制多个饼状图....................167	
5.1.7	绘制多系列饼状图................171	
5.1.8	绘制圈状饼状图....................171	
5.1.9	绘制环状饼状图....................172	
5.1.10	绘制半饼状图......................173	
5.1.11	使用库 pandas、numpy 和 matplotlib 绘制饼状图.....174	

5.2 爬取热门电影信息并制作可视化分析饼状图..174
　　5.2.1 创建 MySQL 数据库.............174
　　5.2.2 爬取并分析电影数据............176
5.3 机器学习实战：Scikit-Learn 聚类分析并可视化处理..179
　　5.3.1 准备饼状图............................179
　　5.3.2 聚类处理................................180
　　5.3.3 生成统计柱状图....................180
5.4 可视化展示名著《西游记》中出现频率最多的文字..182
　　5.4.1 单元测试文件........................182
　　5.4.2 GUI 界面...............................182
　　5.4.3 设置所需显示的出现频率.....185

第 6 章　绘制其他图形..........................187

6.1 绘制雷达图..188
　　6.1.1 创建极坐标图........................188
　　6.1.2 设置极坐标的正方向............188
　　6.1.3 绘制一个基本的雷达图........189
　　6.1.4 绘制 XX 战队 2020 绝地求生战绩的雷达图...........................190
　　6.1.5 使用雷达图比较两名研发部同事的能力............................191
　　6.1.6 绘制汽车性能雷达图............192
　　6.1.7 使用 pygal 绘制雷达图.........193
　　6.1.8 绘制主流编程语言的雷达图................................194
6.2 绘制热力图..195
　　6.2.1 绘制热力图的函数................195
　　6.2.2 绘制一个简单的热力图........197
　　6.2.3 使用库 matplotlib 绘制热力图................................197
6.3 将 Excel 文件中的地址信息可视化为交通热力图..199
　　6.3.1 将地址转换为 JS 格式.........199
　　6.3.2 将 JS 地址转换为坐标..........199
　　6.3.3 在地图中显示地址的热力信息................................201
6.4 使用热点图可视化展示电视剧的收视率..202
　　6.4.1 爬虫爬取电视剧资料............202
　　6.4.2 使用热点图实现可视化........204
6.5 行人重识别并绘制行走热力图............207
　　6.5.1 安装第三方库 pytorch..........208
　　6.5.2 编写识别程序和绘图程序....208
6.6 绘制词云图..210
　　6.6.1 绘制 B 站词云图..................210
　　6.6.2 绘制知乎词云图...................211
6.7 使用热力图可视化展示某城市的房价信息..212
　　6.7.1 准备数据................................212
　　6.7.2 使用热力图可视化展示信息................................213

第 7 章　商业应用：电影票房数据可视化..217

7.1 需求分析..218
7.2 模块架构..218
7.3 爬虫抓取数据..219
　　7.3.1 分析网页................................219
　　7.3.2 破解反爬................................220

    7.3.3 构造请求头 223
    7.3.4 实现具体爬虫功能 224
    7.3.5 将爬取的信息保存到
          数据库 226
  7.4 数据可视化分析 227
    7.4.1 电影票房 TOP10 227
    7.4.2 电影评分 TOP10 229
    7.4.3 电影人气 TOP10 230
    7.4.4 每月电影上映数量 231
    7.4.5 每月电影票房 233
    7.4.6 中外票房对比 234
    7.4.7 名利双收 TOP10 236
    7.4.8 叫座不叫好 TOP10 237
    7.4.9 电影类型分布 238

第 8 章 商业应用：房地产市场数据可视化 241

  8.1 背景介绍 242
  8.2 需求分析 242
  8.3 模块架构 242
  8.4 系统设置 243
    8.4.1 选择版本 243
    8.4.2 保存日志信息 244
    8.4.3 设置创建的文件名 244
    8.4.4 设置抓取城市 245
    8.4.5 处理区县信息 247
  8.5 破解反爬机制 249
    8.5.1 定义爬虫基类 249
    8.5.2 浏览器用户代理 250
    8.5.3 在线 IP 代理 251
  8.6 爬虫抓取信息 251
    8.6.1 设置解析元素 251
    8.6.2 爬取二手房信息 252
    8.6.3 爬取楼盘信息 255
    8.6.4 爬取小区信息 258
    8.6.5 抓取租房信息 262
  8.7 数据可视化 266

    8.7.1 爬取数据并保存到数据库 266
    8.7.2 可视化济南市房价最贵的
          4 个小区 270
    8.7.3 可视化济南市主要地区的房价
          均价 271
    8.7.4 可视化济南市主要地区的房源
          数量 272
    8.7.5 可视化济南市各区的房源数量
          所占百分比 274

第 9 章 商业应用：交通数据可视化 277

  9.1 系统架构分析 278
  9.2 从 CSV 文件读取数据 278
    9.2.1 读取显示 CSV 文件中的前 3 条
          骑行数据 278
    9.2.2 读取显示 CSV 文件中指定列的
          数据 283
    9.2.3 用统计图可视化 CSV 文件中的
          数据 283
    9.2.4 选择指定数据 284
  9.3 日期相关操作 290
    9.3.1 统计每个月的骑行数据 290
    9.3.2 展示某街道前 5 天的骑行数据
          信息 291
    9.3.3 统计周一到周日每天的
          数据 292
    9.3.4 使用 matplotlib 图表可视化
          展示统计数据 293

第 10 章 商业应用：招聘信息可视化 297

  10.1 系统背景介绍 298
  10.2 系统架构分析 298
  10.3 系统设置 299
  10.4 网络爬虫 299
    10.4.1 建立和数据库的连接 300
    10.4.2 设置 HTTP 请求头
           User-Agent 300

10.4.3　抓取信息.................................301
　　10.4.4　将抓取的信息添加到
　　　　　　数据库.................................302
　　10.4.5　处理薪资数据.....................303
　　10.4.6　清空数据库数据.................304
　　10.4.7　执行爬虫程序.....................304
10.5　信息分离统计.....................................304
　　10.5.1　根据"工作经验"分析
　　　　　　数据.....................................305
　　10.5.2　根据"工作地区"分析
　　　　　　数据.....................................306

　　10.5.3　根据"薪资水平"分析
　　　　　　数据.....................................307
　　10.5.4　根据"学历水平"分析
　　　　　　数据.....................................308
10.6　数据可视化...309
　　10.6.1　Flask Web 架构.....................309
　　10.6.2　Web 主页..............................311
　　10.6.3　数据展示页面.....................312
　　10.6.4　数据可视化页面.................314

扫码可下载全书案例源代码

第 1 章

数据采集

进行数据可视化处理的前提是有要处理的数据,这些数据是从哪里来的呢?一些是个人、机构或组织提供的,一些需要大家从网络中收集归纳。本章将详细讲解使用 Python 语言采集数据的知识,为读者步入本书后面知识的学习打下基础。

1.1 处理网络数据

互联网改变了人们的生活方式，极大地方便了人们生活中的方方面面，本书要分析的数据绝大多数来自网络。在本节的内容中，将详细讲解使用网络库处理网络数据的基本知识。

扫码观看本节视频讲解

1.1.1 解析 HTML 和 XML 数据

1. 使用内置库解析 XML 数据

在 Python 应用程序中，有两种常见的 XML 编程接口，分别是 SAX 和 DOM。所以，与之对应的是 Python 语言有两种解析 XML 文件的方法，分别是 SAX 和 DOM 方法。其中库 xml 由以下核心模块构成。

- xml.etree.elementtree：提供了处理 ElementTree 的成员，是一个轻量级的 XML 处理器。
- xml.dom：用于定义 DOM 的 API，提供了处理 DOM 标记的成员。
- xml.dom.minidom：提供了处理最小 DOM 的成员。
- xml.dom.pulldom：提供了构建部分 DOM 树的成员。
- xml.sax：提供了处理 SAX2 的基类和方法成员。
- xml.parsers.expat：绑定了 Expat 解析器的功能，能够使用注册处理器处理不同 XML 文档部分。

例如，在下面的实例代码中，演示了使用库 xml.etree.elementtree 读取 XML 文件的过程。其中 XML 文件 test.xml 的具体实现代码如下。

> 源码路径：codes\1\1-1\test.xml

```
<students>
   <student name='赵敏' sex='男' age='35'/>
   <student name='周芷若' sex='男' age='38'/>
   <student name='小昭' sex='女' age='22'/>
</students>
```

例如，在下面的实例代码中，演示了使用 SAX 方法解析 XML 文件的过程。其中实例文件 movies.xml 是一个基本的 XML 文件，里面保存着一些和电影有关的资料信息。文件 movies.xml 的具体实现代码如下。

> 源码路径：codes\1\1-1\movies.xml

```
<collection shelf="Root">
<movie title="深入敌后">
<type>War, Thriller</type>
<format>DVD</format>
<year>2003</year>
```

```
<rating>三星</rating>
<stars>10</stars>
<description>战争故事</description>
</movie>
<movie title="变形金刚">
<type>Anime, Science Fiction</type>
<format>DVD</format>
<year>1989</year>
<rating>五星</rating>
<stars>8</stars>
<description>科幻片</description>
</movie>
<movie title="枪神">
<type>Anime, Action</type>
<format>DVD</format>
<episodes>4</episodes>
<rating>四星</rating>
<stars>10</stars>
<description>警匪片</description>
</movie>
<movie title="伊师塔">
<type>Comedy</type>
<format>VHS</format>
<rating>五星</rating>
<stars>2</stars>
<description>希腊神话</description>
</movie>
</collection>
```

实例文件 sax.py 的功能是解析文件 movies.xml 的内容，具体实现代码如下。

源码路径：**codes\1\1-1\sax.py**

```python
import xml.sax
class MovieHandler( xml.sax.ContentHandler ):
def __init__(self):
    self.CurrentData = ""
    self.type = ""
    self.format = ""
    self.year = ""
    self.rating = ""
    self.stars = ""
    self.description = ""
   # 元素开始调用
def startElement(self, tag, attributes):
    self.CurrentData = tag
if tag == "movie":
print ("*****Movie*****")
title = attributes["title"]
print ("Title:", title)
   # 元素结束调用
def endElement(self, tag):
    if self.CurrentData == "type":           #处理 XML 中的 type 元素
print ("Type:", self.type)
    elif self.CurrentData == "format":       #处理 XML 中的 format 元素
```

```
        print ("Format:", self.format)
            elif self.CurrentData == "year":              #处理 XML 中的 year 元素
        print ("Year:", self.year)
            elif self.CurrentData == "rating":            #处理 XML 中的 rating 元素
        print ("Rating:", self.rating)
            elif self.CurrentData == "stars":             #处理 XML 中的 stars 元素
        print ("Stars:", self.stars)
            elif self.CurrentData == "description":       #处理 XML 中的 description 元素
        print ("Description:", self.description)
            self.CurrentData = ""
        # 读取字符时调用
        def characters(self, content):
    if self.CurrentData == "type":
            self.type = content
    elif self.CurrentData == "format":
            self.format = content
    elif self.CurrentData == "year":
            self.year = content
    elif self.CurrentData == "rating":
            self.rating = content
    elif self.CurrentData == "stars":
            self.stars = content
    elif self.CurrentData == "description":
            self.description = content
if ( __name__ == "__main__"):
    # 创建一个 XMLReader
parser = xml.sax.make_parser()
    # turn off namespaces
parser.setFeature(xml.sax.handler.feature_namespaces, 0)
    # 重写 ContentHandler
    Handler = MovieHandler()
parser.setContentHandler( Handler )
parser.parse("movies.xml")
```

执行后的效果如图 1-1 所示。

```
*****Movie*****
Title: 深入敌后
Type: War, Thriller
Format: DVD
Year: 2003
Rating: 三星
Stars: 10
Description: 战争故事
*****Movie*****
Title: 变形金刚
Type: Anime, Science Fiction
Format: DVD
Year: 1989
Rating: 五星
Stars: 8
Description: 科幻片
*****Movie*****
Title: 枪神
Type: Anime, Action
Format: DVD
Rating: 四星
Stars: 10
Description: 警匪片
```

图 1-1　执行后的效果

例如，在下面的实例文件 dom.py 中，演示了使用 DOM 方法解析 XML 文件的过程。实例文件 dom.py 的功能是解析文件 movies.xml 的内容，具体实现代码如下。

源码路径：codes\1\1-1\dom.py

```python
from xml.dom.minidom import parse
import xml.dom.minidom
# 使用minidom.parse解析器解析 XML 文件
DOMTree = xml.dom.minidom.parse("movies.xml")
collection = DOMTree.documentElement
if collection.hasAttribute("shelf"):
   print ("根元素是 : %s" % collection.getAttribute("shelf"))

# 在集合中获取所有电影
movies = collection.getElementsByTagName("movie")

# 打印每部电影的详细信息
for movie in movies:
   print ("*****Movie*****")
   if movie.hasAttribute("title"):
      print ("Title 电影名: %s" % movie.getAttribute("title"))

   type = movie.getElementsByTagName('type')[0]
   print ("Type 电影类型: %s" % type.childNodes[0].data)
   format = movie.getElementsByTagName('format')[0]
   print ("Format 电影格式: %s" % format.childNodes[0].data)
   rating = movie.getElementsByTagName('rating')[0]
   print ("Rating 电影评分: %s" % rating.childNodes[0].data)
   description = movie.getElementsByTagName('description')[0]
   print ("Description 电影简介: %s" % description.childNodes[0].data)
```

执行后会输出：

```
根元素是 : Root
*****Movie*****
Title 电影名：深入敌后
Type 电影类型: War, Thriller
Format 电影格式：DVD
Rating 电影评分：三星
Description 电影简介：战争故事
*****Movie*****
Title 电影名：变形金刚
Type 电影类型: Anime, Science Fiction
Format 电影格式：DVD
Rating 电影评分：五星
Description 电影简介：科幻片
*****Movie*****
Title 电影名：枪神
Type 电影类型: Anime, Action
Format 电影格式：DVD
Rating 电影评分：四星
Description 电影简介：警匪片
*****Movie*****
Title 电影名：伊师塔
```

```
Type 电影类型：Comedy
Format 电影格式：VHS
Rating 电影评分：五星
Description 电影简介：希腊神话
```

2. 使用库 Beautiful Soup 解析网络数据

Beautiful Soup 是一个重要的 Python 库，其功能是将 HTML 和 XML 文件的标签信息解析成树状结构，然后提取 HTML 或 XML 文件中指定标签属性对应的数据。库 Beautiful Soup 经常被用在爬虫项目中，通过使用库 Beautiful Soup 可以大大提高开发效率。

Beautiful Soup 3 目前已经停止开发，其官方推荐使用 Beautiful Soup 4，本书讲解的是 Beautiful Soup 4。开发者可以使用以下两种命令安装库 Beautiful Soup：

```
pip install beautifulsoup4
easy_install beautifulsoup4
```

在安装 Beautiful Soup 4 后还需要安装文件解析器，Beautiful Soup 不但支持 Python 标准库中的 HTML 解析器，而且还支持第三方的解析器(如 lxml)。根据开发者所用操作系统的不同，可以使用以下命令来安装 lxml：

```
$ apt-get install Python-lxml
$ easy_install lxml
$ pip install lxml
```

例如，在下面的实例文件 bs01.py 中，演示了使用库 Beautiful Soup 解析 HTML 代码的过程。

源码路径：codes\1\1-1\bs01.py

```
from bs4 import BeautifulSoup
html_doc = """
<html><head><title>The Dormouse's story</title></head>
<body>
<p class="title"><b>睡鼠的故事</b></p>

<p class="story">在很久以前有三个可爱的小熊宝宝，名字分别是
<a href="http://example.com/elsie" class="sister" id="link1">Elsie</a>,
<a href="http://example.com/lacie" class="sister" id="link2">Lacie</a>和
<a href="http://example.com/tillie" class="sister" id="link3">Tillie</a>;
and they lived at the bottom of a well.</p>

<p class="story">...</p>
"""
soup = BeautifulSoup(html_doc,"lxml")
print(soup)
```

通过上述代码，解析了 html_doc 中的 HTML 代码，执行后会输出解析结果：

```
<html><head><title>The Dormouse's story</title></head>
<body>
<p class="title"><b>睡鼠的故事</b></p>
<p class="story">在很久以前有三个可爱的小熊宝宝，名字分别是
```

```
<a class="sister" href="http://example.com/elsie" id="link1">Elsie</a>,
<a class="sister" href="http://example.com/lacie" id="link2">Lacie</a>和
<a class="sister" href="http://example.com/tillie" id="link3">Tillie</a>;
and they lived at the bottom of a well.</p>
<p class="story">...</p>
</body></html>
```

在下面的实例文件 bs02.py 中,演示了使用库 Beautiful Soup 解析指定 HTML 标签的过程。

源码路径:codes\1\1-1\bs02.py

```python
from bs4 import BeautifulSoup

html = '''
<html><head><title>睡鼠的故事</title></head>
<body>
<p class="title"><b>睡鼠的故事</b></p>

<p class="story">0 在很久以前有三个可爱的小熊宝宝,名字分别是
<a href="http://example.com/elsie" class="sister" id="link1">Elsie</a>,
<a href="http://example.com/lacie" class="sister" id="link2">Lacie</a> 和
<a href="http://example.com/tillie" class="sister" id="link3">Tillie</a>;
他们快乐的住在大森林里</p>
<p class="story">...</p>
'''
soup = BeautifulSoup(html,'lxml')
print(soup.title)
print(soup.title.name)
print(soup.title.string)
print(soup.title.parent.name)
print(soup.p)
print(soup.p["class"])
print(soup.a)
print(soup.find_all('a'))
print(soup.find(id='link3'))
```

执行后将输出指定标签的信息:

```
<title>睡鼠的故事</title>
title
睡鼠的故事
head
<p class="title"><b>睡鼠的故事</b></p>
['title']
<a class="sister" href="http://example.com/elsie" id="link1">Elsie</a>
[<a class="sister" href="http://example.com/elsie" id="link1">Elsie</a>, <a class="sister" href="http://example.com/lacie" id="link2">Lacie</a>, <a class="sister" href="http://example.com/tillie" id="link3">Tillie</a>]
<a class="sister" href="http://example.com/tillie" id="link3">Tillie</a>
```

3. 使用库 bleach 过滤数据

在使用 Python 开发 Web 程序时,开发者面临一个十分重要的安全性问题,即跨站脚本

注入攻击(黑客利用网站漏洞从用户一端盗取重要信息)。为了解决跨站脚本注入攻击漏洞，最常用的做法是设置一个访问白名单，设置只显示指定的 HTML 标签和属性。在现实应用中，最常用的 HTML 过滤库是 Bleach，能够实现基于白名单的 HTML 清理和文本链接模块。

可以使用以下两种命令安装库 bleach：

```
pip install bleach
easy_install bleach
```

例如，在下面的实例文件 guolv.py 中，演示了使用方法 bleach.clean( )过滤处理 HTML 标签的过程。

源码路径：codes\1\1-1\guolv.py

```python
import bleach
# tag 参数示例
print(bleach.clean(
    u'<b><i>例子1</i></b>',
    tags=['b'],
))

# attributes 为 list 示例
print(bleach.clean(
    u'<p class="foo" style="color: red; font-weight: bold;">例子2</p>',
    tags=['p'],
    attributes=['style'],
    styles=['color'],
))
# attributes 为 dict 示例
attrs = {
    '*': ['class'],
    'a': ['href', 'rel'],
    'img': ['alt'],
}
print(bleach.clean(
    u'<img alt="an example" width=500>例子3',
    tags=['img'],
    attributes=attrs
))

# attributes 为 function 示例
def allow_h(tag, name, value):
    return name[0] == 'h'
print(bleach.clean(
    u'<a href="http://example.com" title="link">例子4</a>',
    tags=['a'],
    attributes=allow_h,
))

# style 参数示例
tags = ['p', 'em', 'strong']
attrs = {
    '*': ['style']
```

```
}
styles = ['color', 'font-weight']
print(bleach.clean(
    u'<p style="font-weight: heavy;">例子 5</p>',
    tags=tags,
    attributes=attrs,
    styles=styles
))
# protocol 参数示例
print(bleach.clean(
    '<a href="smb://more_text">例子 6</a>',
    protocols=['http', 'https', 'smb']
))
print(bleach.clean(
    '<a href="smb://more_text">例子 7</a>',
    protocols=bleach.ALLOWED_PROTOCOLS + ['smb']
))

#strip 参数示例
print(bleach.clean('<span>例子 8</span>'))
print(bleach.clean('<b><span>例子 9</span></b>', tags=['b']))

print(bleach.clean('<span>例子 10</span>', strip=True))
print(bleach.clean('<b><span>例子 11</span></b>', tags=['b'], strip=True))

# strip_comments 参数示例
html = 'my<!-- commented --> html'
print(bleach.clean(html))
print(bleach.clean(html, strip_comments=False))
```

执行后会输出：

```
<b>&lt;i&gt;例子 1&lt;/i&gt;</b>
<p style="color: red;">例子 2</p>
<img alt="an example">例子 3
<a href="http://example.com">例子 4</a>
<p style="font-weight: heavy;">例子 5</p>
<a href="smb://more_text">例子 6</a>
<a href="smb://more_text">例子 7</a>
&lt;span&gt;例子 8&lt;/span&gt;
<b>&lt;span&gt;例子 9&lt;/span&gt;</b>
例子 10
<b>例子 11</b>
my html
my<!-- commented --> html
```

4. 使用库 html5lib 解析网络数据

在 Python 程序中，可以使用库 html5lib 解析 HTML 文件。库 html5lib 是用纯 Python 语言编写实现的，其解析方式与浏览器相同。在本章前面讲解的库 Beautiful Soup，使用的是 lxml 解析器，而库 html5lib 是 Beautiful Soup 支持的另一种解析器。

可以使用以下两种命令安装库 html5lib：

```
pip install html5lib
easy_install html5lib
```

例如，在下面的实例文件 ht501.py 中，演示了使用 html5lib 解析 HTML 文件的过程。

源码路径：codes\1\1-1\ht501.py

```
from bs4 import BeautifulSoup
html_doc = """
<html><head><title>睡鼠的故事</title></head>
<body>
<p class="title"><b>睡鼠的故事</b></p>

<p class="story">在很久以前有三个可爱的小熊宝宝，名字分别是
<a href="http://example.com/elsie" class="sister" id="link1">Elsie</a>,
<a href="http://example.com/lacie" class="sister" id="link2">Lacie</a> 和
<a href="http://example.com/tillie" class="sister" id="link3">Tillie</a>;
他们快乐的生活在大森林里.</p>

<p class="story">...</p>
"""
soup = BeautifulSoup(html_doc,"html5lib")
print(soup)
```

执行后会输出：

```
<html><head><title>睡鼠的故事</title></head>
<body>
<p class="title"><b>睡鼠的故事</b></p>

<p class="story">在很久以前有三个可爱的小熊宝宝，名字分别是
<a class="sister" href="http://example.com/elsie" id="link1">Elsie</a>,
<a class="sister" href="http://example.com/lacie" id="link2">Lacie</a>和
<a class="sister" href="http://example.com/tillie" id="link3">Tillie</a>;
他们快乐的生活在大森林里.</p>

<p class="story">...</p>
</body></html>
```

5. 使用库 MarkupSafe 解析数据

在 Python 程序中，使用库 MarkupSafe 可以将具有特殊含义的字符替换掉，这样可以减轻注入攻击(把用户输入、提交的数据当作代码来执行)风险，能够将不受信任用户输入的信息安全地显示在页面上。

可以使用以下两种命令安装库 MarkupSafe：

```
pip install MarkupSafe
easy_install MarkupSafe
```

例如，在下面的实例文件 mark01.py 中，演示了使用库 MarkupSafe 构建安全 HTML 的过程。

源码路径：codes\1\1-1\mark01.py

```
from markupsafe import Markup, escape
#实现支持 HTML 字符串的 Unicode 子类
print(escape("<script>alert(document.cookie);</script>"))
tmpl = Markup("<em>%s</em>")
print(tmpl % "Peter > Lustig")

#可以通过重写__html__功能自定义等效 HTML 标记
class Foo(object):
    def __html__(self):
        return '<strong>Nice</strong>'

print(escape(Foo()))
print(Markup(Foo()))
```

执行后会输出：

```
&lt;script&gt;alert(document.cookie);&lt;/script&gt;
<em>Peter &gt; Lustig</em>
<strong>Nice</strong>
<strong>Nice</strong>
```

1.1.2 处理 HTTP 数据

HTTP(HyperText Transfer Protocol)是互联网上应用最广泛的一种网络协议，所有的 WWW 文件都必须遵守这个标准。在下面的内容中，将详细讲解常用的处理 HTTP 数据的知识。

1. 使用 Python 内置库 http 处理网络数据

在 Python 语言中，使用内置库 http 实现了对 HTTP 协议的封装。在库 http 中主要包含以下模块。

- http.client：底层的 HTTP 协议客户端，可以为 urllib.request 模块所用。
- http.server：提供了处理 socketserver 模块的功能类。
- http.cookies：提供了在 HTTP 传输过程中处理 Cookies 应用的功能类。
- http.cookiejar：提供了实现 Cookies 持久化支持的功能类。

在 http.client 模块中，主要包括以下两个处理客户端应用的类。

- HTTPConnection：基于 HTTP 协议的访问客户端。
- HTTPResponse：基于 HTTP 协议的服务端回应。

例如，在下面的实例文件 fang.py 中，演示了使用 http.client.HTTPConnection 对象访问指定网站的过程。

源码路径：codes\1\1-1\fang.py

```
from http.client import HTTPConnection    #导入内置模块
#基于 HTTP 协议的访问客户端
mc = HTTPConnection('www.baidu.com:80')
```

```
mc.request('GET','/')                          #设置 GET 请求方法
res = mc.getresponse()                         #获取访问的网页
print(res.status,res.reason)                   #打印输出响应的状态
print(res.read().decode('utf-8'))              #显示获取的内容
```

在上述实例代码中只是实现了一个基本的访问实例,首先实例化 http.client.HTTPConnection 对指定请求的方法为 GET,然后使用 getresponse( )方法获取访问的网页,并打印输出响应的状态。执行效果如图 1-2 所示。

```
200 OK
<!DOCTYPE html><!--STATUS OK-->
<html>
    <head>
        <meta http-equiv="content-type" content="text/html;charset=utf-8">
        <meta http-equiv="X-UA-Compatible" content="IE=Edge">
        <link rel="dns-prefetch" href="//s1.bdstatic.com"/>
        <link rel="dns-prefetch" href="//t1.baidu.com"/>
        <link rel="dns-prefetch" href="//t2.baidu.com"/>
        <link rel="dns-prefetch" href="//t3.baidu.com"/>
        <link rel="dns-prefetch" href="//t10.baidu.com"/>
        <link rel="dns-prefetch" href="//t11.baidu.com"/>
        <link rel="dns-prefetch" href="//t12.baidu.com"/>
        <link rel="dns-prefetch" href="//b1.bdstatic.com"/>
        <title>百度一下,你就知道</title>
        <link href="http://s1.bdstatic.com/r/www/cache/static/home/css/index.css" rel="stylesheet" type="text/css" />
        <!--[if lte IE 8]><style index="index" >#content{height:480px\9}#m{top:260px\9}</style><![endif]-->
        <!--[if IE 8]><style index="index" >ul a.mnav,ul a.mnav:visited{font-family:simsun}</style><![endif]-->
        <script>var hashMatch = document.location.href.match(/\$+(.*wd=[`&].+)/);if (hashMatch && hashMatch[0] && hashMatch[1])
        <script>function h(obj){obj.style.behavior='url(#default#homepage)';var a = obj.setHomePage('//www.baidu.com');};</scri
```

图 1-2 执行效果

在现实应用中,有时需要通过 HTTP 协议以客户端的形式访问多种服务,如下载服务器中的数据或同一个基于 REST 的 API 进行交互。通过使用 urllib.request 模块,可以实现简单的客户端访问任务。例如,要发送一个简单的 HTTP GET 请求到远端服务器上,只需通过下面的实例文件 fang1.py 即可实现。

源码路径:codes\1\1-1\fang1.py

```python
from urllib import request, parse

# Base URL being accessed
url = 'http://httpbin.org/get'

# Dictionary of query parameters (if any)
parms = {
   'name1' : 'value1',
   'name2' : 'value2'
}

# Encode the query string
querystring = parse.urlencode(parms)

# Make a GET request and read the response
u = request.urlopen(url+'?' + querystring)
resp = u.read()

import json
from pprint import pprint
```

```
json_resp = json.loads(resp.decode('utf-8'))
pprint(json_resp)
```

执行后会输出:

```
{'args': {'name1': 'value1', 'name2': 'value2'},
 'headers': {'Accept-Encoding': 'identity',
             'Connection': 'close',
             'Host': 'httpbin.org',
             'User-Agent': 'Python-urllib/3.6'},
 'origin': '27.211.158.101',
 'url': 'http://httpbin.org/get?name1=value1&name2=value2'}
```

2. 使用库 requests 处理网络数据

库 requests 是使用 Python 语言对库 urllib 的封装升级，使用库 requests 会比使用 urllib 更加方便。可以使用以下两种命令安装库 requests：

```
pip install requests
easy_install requests
```

例如，在下面的实例文件 Requests01.py 中，演示了使用库 requests 返回指定 URL 地址请求的过程。

源码路径：codes\1\1-1\Requests01.py

```python
import requests

r = requests.get(url='http://www.toppr.net')    # 最基本的 GET 请求
print(r.status_code)    # 获取返回状态
r = requests.get(url='http://www.toppr.net', params={'wd': 'python'})  # 带参数的 GET 请求
print(r.url)
print(r.text)    # 打印解码后的返回数据
```

在上述代码中，创建了一个名为 r 的 Response 对象，可以从这个对象中获取所有想要的信息。执行后会输出：

```
200
http://www.toppr.net/?wd=python
<!DOCTYPE html PUBLIC "-//W3C//DTD XHTML 1.0 Transitional//EN"
"http://www.w3.org/TR/xhtml1/DTD/xhtml1-transitional.dtd">
<html xmlns="http://www.w3.org/1999/xhtml">
<head>
<meta http-equiv="X-UA-Compatible" content="IE=edge">
<meta http-equiv="Content-Type" content="text/html; charset=gbk" />
<title>门户 - Powered by Discuz!</title>

<meta name="keywords" content="门户" />
<meta name="description" content="门户 " />
<meta name="generator" content="Discuz! X3.2" />
#省略后面的结果
```

上述实例只是演示了 get 接口的用法，其实其他接口的用法也十分简单：

```
requests.get('https://github.com/timeline.json') #GET 请求
requests.post("http://httpbin.org/post") #POST 请求
requests.put("http://httpbin.org/put") #PUT 请求
requests.delete("http://httpbin.org/delete") #DELETE 请求
requests.head("http://httpbin.org/get") #HEAD 请求
requests.options("http://httpbin.org/get") #OPTIONS 请求
```

例如，想查询 http://httpbin.org/get 页面的具体参数，需要在 url 里面加上这个参数。假如想看有没有 Host=httpbin.org 这条数据，url 形式应该是 http://httpbin.org/get?Host=httpbin.org。例如，在下面的实例文件 Requests02.py 中，提交的数据是往这个地址传送 data 里面的数据。

源码路径：codes\1\1-1\Requests02.py

```python
import requests
url = 'http://httpbin.org/get'
data = {
    'name': 'python',
    'age': '25'
}
response = requests.get(url, params=data)
print(response.url)
print(response.text)
```

执行后会输出：

```
http://httpbin.org/get?name=zhangsan&age=25
{
  "args": {
    "age": "25",
    "name": "python "
  },
  "headers": {
    "Accept": "*/*",
    "Accept-Encoding": "gzip, deflate",
    "Connection": "close",
    "Host": "httpbin.org",
    "User-Agent": "python-requests/2.12.4"
  },
  "origin": "39.71.61.153",
  "url": "http://httpbin.org/get?name=python&age=25"
}
```

3. 使用库 urllib3 处理网络数据

在 Python 应用中，库 urllib3 提供了一个线程安全的连接池，能够以 post 方式传输文件。可以使用以下两种命令安装库 urllib3：

```
pip install urllib3
easy_install urllib3
```

在下面的实例文件 urllib303.py 中，演示了使用库 urllib3 中的 post( )方法创建请求的过程。

源码路径：codes\1\1-1\urllib303.py

```python
import urllib3
http = urllib3.PoolManager()
#信息头
header = {
      'User-Agent': 'Mozilla/5.0 (Windows NT 2.1; Win64; x64) AppleWebKit/537.36 (KHTML, like Gecko) Chrome/63.0.3239.108 Safari/537.36'
}
r = http.request('POST',
              'http://httpbin.org/post',
              fields={'hello':'Python'},
              headers=header)
print(r.data.decode())

# 对于 POST 和 PUT 请求(request)，需要手动对传入数据进行编码，然后加在 URL 之后：
encode_arg = urllib.parse.urlencode({'arg': '我的信息：'})
print(encode_arg.encode())
r = http.request('POST',
              'http://httpbin.org/post?'+encode_arg,
              headers=header)
# unicode 解码
print(r.data.decode('unicode_escape'))
```

执行后会输出：

```
{
  "args": {},
  "data": "",
  "files": {},
  "form": {
    "hello": "Python"
  },
  "headers": {
    "Accept-Encoding": "identity",
    "Connection": "close",
    "Content-Length": "129",
    "Content-Type": "multipart/form-data; boundary=b33b20053e6444ee947a6b7b3f4572b2",
    "Host": "httpbin.org",
    "User-Agent": "Mozilla/5.0 (Windows NT 2.1; Win64; x64) AppleWebKit/537.36 (KHTML, like Gecko) Chrome/63.0.3239.108 Safari/537.36"
  },
  "json": null,
  "origin": "39.71.61.153",
  "url": "http://httpbin.org/post"
}

b'arg=%E6%88%91%E7%9A%84'
{
  "args": {
    "arg": "我的"
  },
  "data": "",
  "files": {},
  "form": {},
```

```
  "headers": {
    "Accept-Encoding": "identity",
    "Connection": "close",
    "Content-Length": "0",
    "Host": "httpbin.org",
    "User-Agent": "Mozilla/5.0 (Windows NT 2.1; Win64; x64) AppleWebKit/537.36
(KHTML, like Gecko) Chrome/63.0.3239.108 Safari/537.36"
  },
  "json": null,
  "origin": "39.71.61.153",
  "url": "http://httpbin.org/post?arg=我的信息"
}
```

例如，在下面的实例文件 urllib305.py 中，演示了使用库 urllib3 获取远程 CSV 数据的过程。

源码路径：codes\1\1-1\urllib305.py

```
import urllib3
#两个文件的源url
url1 =
'http://earthquake.usgs.gov/earthquakes/feed/v1.0/summary/all_week.csv'
url2 =
'http://earthquake.usgs.gov/earthquakes/feed/v1.0/summary/all_month.csv'
#开始创建一个HTTP连接池
http = urllib3.PoolManager()
#请求第一个文件并将结果写入文件：
response = http.request('GET', url1)
with open('all_week.csv', 'wb') as f:
    f.write(response.data)
#请求第二个文件并将结果写入CSV文件：
response = http.request('GET', url2)
with open('all_month.csv', 'wb') as f:
    f.write(response.data)

#最后释放HTTP连接的占用资源：
response.release_conn()
```

执行后会将这两个远程 CSV 文件下载保存到本地，如图 1-3 所示。

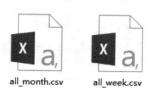

图 1-3　下载保存到本地的 CSV 文件

在下面的实例文件 urllib302.py 中，演示了使用库 urllib3 抓取显示凤凰网头条新闻的方法。

源码路径：codes\1\1-1\urllib302.py

```
from bs4 import BeautifulSoup
import urllib3
```

```python
def get_html(url):
    try:
        userAgent = 'Mozilla/5.0 (Windows; U; Windows NT 2.1; en-US; rv:1.9.1.6) Gecko/20091201 Firefox/3.5.6'
        http = urllib3.PoolManager(timeout=2)
        response = http.request('get', url, headers={'User_Agent': userAgent})
        html = response.data
        return html
    except Exception as e:
        print(e)
        return None

def get_soup(url):
    if not url:
        return None
    try:
        soup = BeautifulSoup(get_html(url))
    except Exception as e:
        print(e)
        return None
    return soup

def get_ele(soup, selector):
    try:
        ele = soup.select(selector)
        return ele
    except Exception as e:
        print(e)
    return None

def main():
    url = 'http://www.ifeng.com/'
    soup = get_soup(url)
    ele = get_ele(soup, '#headLineDefault > ul > ul:nth-of-type(1) > li.topNews > h1 > a')
    headline = ele[0].text.strip()
    print(headline)

if __name__ == '__main__':
    main()
```

因为头条新闻是随着时间的推移发生变化的，所以每次的执行效果可能不一样。

1.1.3 处理 URL 数据

URL(Uniform Resource Locator, 统一资源定位器)也就是平常所说的 WWW 网址。下面将详细讲解使用 Python 语言处理 URL 数据的知识。

1. 使用内置库 urllib

在 Python 程序中，可以内置库 urllib 处理 URL 请求。urllib 非常简单而且易用，在内置库 urllib 中主要包括以下几个模块。

- urllib.request：用于打开指定的 URL 网址。
- urllib.error：用于处理 URL 访问异常。
- urllib.parse：用于解析指定的 URL。
- urllib.robotparser：用于解析 robots.txt 文件，robots 是 Web 网站跟爬虫之间的协议，可以用 txt 格式的文本方式告诉对应的爬虫被允许的权限。

(1) 使用 urllib.request 模块。

在 urllib.request 模块中定义了打开指定 URL 的方法和类，甚至可以实现身份验证、URL 重定向和 Cookies 存储等功能。例如，在下面的实例文件 url.py 中，演示了使用方法 urlopen( ) 在百度搜索关键词中得到第一页链接的过程。

源码路径：codes\1\1-1\url.py

```python
from urllib.request import urlopen       #导入 Python 的内置模块
from urllib.parse import urlencode       #导入 Python 的内置模块
import re                                #导入 Python 的内置模块
##wd = input('输入一个要搜索的关键字：')
wd= 'www.toppr.net'                      #初始化变量 wd
wd = urlencode({'wd':wd})                #对 URL 进行编码
url = 'http://www.baidu.com/s?' + wd     #初始化 url 变量
page = urlopen(url).read()               #打开变量 url 的网页并读取内容
#定义变量 content，对网页进行编码处理，并实现特殊字符处理
content = (page.decode('utf-8')).replace("\n","").replace("\t","")
title = re.findall(r'<h3 class="t".*?h3>', content)
#正则表达式处理
title = [item[item.find('href =')+6:item.find('target=')] for item in title]
#正则表达式处理
title = [item.replace(' ','').replace('"','') for item in title] #正则表达式处理
for item in title:                       #遍历 title
    print(item)                          #打印显示遍历值
```

在上述实例代码中，使用方法 urlencode( ) 对搜索的关键字 www.toppr.net 进行 URL 编码，在拼接到百度的网址后，使用 urlopen( ) 方法发出访问请求并取得结果，最后通过将结果进行解码和正则搜索与字符串处理后输出。如果将程序中的注释去除而把其后一句注释掉，就可以在运行时自主输入搜索的关键词。执行效果如图 1-4 所示。

```
http://www.baidu.com/link?url=hm6N8CdYPCSxsCsreajusLxba8mRVPAgc1D_WBhkYb7
http://www.baidu.com/link?url=Nlf7T18nlQ0pke8pH8CIzg0V_wiqTKRtQ2NXLs-wUzyLHM0UknbUflsJT3DLE2G0m6JW5GlRoBx-GbF6epS7sa
http://www.baidu.com/link?url=cbZcgLHZSTFBp6tFwWGwTuVq6xE3FciM_d-cIH5qRNrkkXaBTLwKKj9n9Rhlvvi8
http://www.baidu.com/link?url=0AHbz_vI3wIC_ocpmRc3jzcjIeu3gDeImuXcGfKulzKGta250-KR-HfGchsHSvGY
http://www.baidu.com/link?url=h591VC_3X6t7hm6eptcTS0dxFe5c4Z7XznyLzpqkJlZ6a0lWptFh4IS37h6LhzIC
```

图 1-4 执行效果

注　意

urllib.response 模块是 urllib 使用的响应类，定义了和 urllib.request 模块类似接口、方法和类，包括 read( ) 和 readline( )。为了节省本书篇幅，本书不再进行讲解。

(2) 使用 urllib.parse 模块。

在 Python 程序中，urllib.parse 模块提供了一些用于处理 URL 字符串的功能。这些功能

主要是通过以下方法实现的。

① 方法 urlparse.urlparse( )。方法 urlparse( )的功能是将 URL 字符串拆分成前面描述的一些主要组件，其语法结构如下：

```
urlparse (urlstr, defProtSch=None, allowFrag=None)
```

方法 urlparse( )将 urlstr 解析成一个 6 元组(prot_sch, net_loc, path, params, query, frag)。如果在 urlstr 中没有提供默认的网络协议或下载方案，defProtSch 会指定一个默认的网络协议。allowFrag 用于标识一个 URL 是否允许使用片段。例如，下面是一个给定 URL 经 urlparse( )后的输出：

```
>>> urlparse.urlparse('http://www.python.org/doc/FAQ.html')
('http', 'www.python.org', '/doc/FAQ.html', '', '', '')
```

② 方法 urlparse.urlunparse( )。方法 urlunparse( )的功能与方法 urlpase( )完全相反，能够将经 urlparse( )处理的 URL 生成 urltup 这个 6 元组(prot_sch, net_loc, path, params, query, frag)，拼接成 URL 并返回。可以用以下方式表示其等价性：

```
urlunparse(urlparse(urlstr)) ≡ urlstr
```

下面是使用 urlunparse( )的语法结构：

```
urlunparse(urltup)
```

③ 方法 urlparse.urljoin( )。在需要处理多个相关的 URL 时需要使用 urljoin( )方法的功能，如在一个 Web 页中可能会产生一系列页面的 URL。方法 urljoin( )的语法格式如下：

```
urljoin (baseurl, newurl, allowFrag=None)
```

方法 urljoin( )能够取得根域名，并将其根路径(net_loc 及其前面的完整路径，但是不包括末端的文件)与 newurl 连接起来。例如，下面的演示过程：

```
>>> urlparse.urljoin('http://www.python.org/doc/FAQ.html',
... 'current/lib/lib.htm')
'http://www.python.org/doc/current/lib/lib.html'
```

假设有一个身份验证(登录名和密码)的 Web 站点，通过验证的最简单方法是在 URL 中使用登录信息进行访问，如 http://username:passwd@www.python.org。但是这种方法的问题是它不具有可编程性。通过使用 urllib 可以很好地解决这个问题，假设合法的登录信息是：

```
LOGIN = 'admin'
PASSWD = "admin"
URL = 'http://localhost'
REALM = 'Secure AAA'
```

此时便可以通过下面的实例文件 pa.py，完成使用 urllib 实现 HTTP 身份验证的过程。

源码路径： **codes\1\1-1\pa.py**

```
import urllib.request, urllib.error, urllib.parse

①LOGIN = 'admin'
PASSWD = "admin"
```

```
  URL = 'http://localhost'
② REALM = 'Secure AAA'

③ def handler_version(url):
     hdlr = urllib.request.HTTPBasicAuthHandler()
     hdlr.add_password(REALM,
         urllib.parse.urlparse(url)[1], LOGIN, PASSWD)
     opener = urllib.request.build_opener(hdlr)
     urllib.request.install_opener(opener)
④    return url

⑤ def request_version(url):
     import base64
     req = urllib.request.Request(url)
     b64str = base64.b64encode(
         bytes('%s:%s' % (LOGIN, PASSWD), 'utf-8'))[:-1]
     req.add_header("Authorization", "Basic %s" % b64str)
⑥    return req

⑦ for funcType in ('handler', 'request'):
     print('*** Using %s:' % funcType.upper())
     url = eval('%s_version' % funcType)(URL)
     f = urllib.request.urlopen(url)
     print(str(f.readline(), 'utf-8'))
⑧ f.close()
```

①~②实现普通的初始化功能，设置合法的登录验证信息。

③~④定义函数 handler_version( )，添加验证信息后建立一个 URL 开启器，安装该开启器以便所有已打开的 URL 都能用到这些验证信息。

⑤~⑥定义函数 request_version( )创建一个 Request 对象，并在 HTTP 请求中添加简单的基 64 编码的验证头信息。在 for 循环里调用 urlopen( )时，该请求用来替换其中的 URL 字符串。

⑦~⑧分别打开了给定的 URL，通过验证后会显示服务器返回的 HTML 页面的第一行(转存了其他行)。如果验证信息无效会返回一个 HTTP 错误(并且不会有 HTML)。

2. 使用库 furl 处理数据

在 Python 应用中，库 furl 是一个快速处理 URL 应用的小型 Python 库，可以使开发者更方便地操作 URL 地址。可以使用以下命令安装 furl：

```
pip install furl
```

例如，在下面的实例文件 url02.py 中，演示了使用库 furl 处理 URL 参数的过程。

源码路径：**codes\1\1-1\url02.py**

```
from furl import furl
f= furl('http://www.baidu.com/?bid=12331')
#打印参数
print(f.args)
#增加参数
```

```
f.args['haha']='123'
print(f.args)
#修改参数
f.args['haha']='124'
print(f.args)
#删除参数
del f.args['haha']
print(f.args)
```

执行后会输出：

```
{'bid': '12331'}
{'bid': '12331', 'haha': '123'}
{'bid': '12331', 'haha': '124'}
{'bid': '12331'}
```

1.2 网络爬虫技术

在 1.1 节的内容中，讲解了处理网络数据的基本知识，其实接下来将要讲解的爬虫技术也属于处理网络数据范畴。本节将详细讲解开发网络爬虫程序的知识，为读者步入本书后面知识的学习打下基础。

扫码观看本节视频讲解

1.2.1 网络爬虫基础

对于网络爬虫这一新奇的概念，大家可以将其理解为在网络中爬行的一只小蜘蛛，如果将互联网比作一张大网，那么可以将爬虫看作在这张网上爬行的蜘蛛，如果它遇到喜欢的资源，那么这只小蜘蛛就会把这些信息爬取下来作为己用。

在浏览网页时可能会看到许多好看的图片，如在打开网页 http://image.baidu.com/时会看到很多张图片及百度搜索框。用肉眼看到的网页实质是由 HTML 代码构成的，爬虫的功能是分析和过滤这些 HTML 代码，然后将有用的资源(如图片和文字等)抓取出来。在现实应用中，被抓取出来的爬虫数据十分重要，通常是进行数据分析的原始资料。

在使用爬虫爬取网络数据时，必须要有一个目标的 URL 才可以获取数据。网络爬虫从一个或若干初始网页的 URL 开始，在爬取网页的过程中，不断从当前页面上抽取新的 URL，并将 URL 放入队列。当满足系统设置的停止条件时，爬虫会停止爬取操作。为了使用抓取到的数据，系统需要存储被爬虫爬取到的网页，然后进行一系列的数据分析和过滤工作。

在现实应用中，网络爬虫获取网络数据的流程如下。

(1) 模拟浏览器发送请求。

在客户端使用 HTTP 技术向目标 Web 页面发起请求，即发送一个 Request。在 Request 中包含请求头和请求体等，是访问目标 Web 页面的前提。Request 请求方式有一个缺陷，即不能执行 JS 和 CSS 代码。

(2) 获取响应内容。

如果目标 Web 服务器能够正常响应，在客户端会得到一个 Response 响应，Response 内容包含 HTML、JSON、图片和视频等类型的信息。

(3) 解析内容。

解析目标网页的内容，既可以使用正则表达式提高解析效率，也可以使用第三方解析库提高解析效率。常用的第三方解析库有 Beautifulsoup 和 pyquery 等。

(4) 保存数据。

在现实应用中，通常将爬取的数据保存到数据库(如 MySQL、Mongdb、Redis 等)或不同格式的文件(如 CSV、JSON 等)中，这样可以为下一步的数据分析工作做好准备。

1.2.2　使用 Beautiful Soup 爬取网络数据

Beautiful Soup 是一个著名的 Python 库，功能是从 HTML 或 XML 文件中提取数据。在现实应用中，通常在爬虫项目中使用 Beautiful Soup 获取网页数据。在作者写作本书时，Beautiful Soup 的最新版本是 Beautiful Soup 4。

1. 安装 Beautiful Soup

在新版的 Debain 或 Ubuntu 系统中，可以通过以下系统软件包管理命令进行安装：

```
$ apt-get install Python-bs4
```

因为 Beautiful Soup 4 是通过 PyPi 发布的，所以可以通过 easy_install 或 pip 来安装。库 Beautiful Soup 4 对应包名是 beautifulsoup4，可以通过以下命令进行安装：

```
easy_install beautifulsoup4
pip install beautifulsoup4
```

在安装 Beautiful Soup 后还需要安装解析器，Beautiful Soup 不但支持 Python 标准库中的 HTML 解析器，而且还支持一些第三方的解析器，如 lxml。根据操作系统不同，可以选择下面的命令来安装 lxml：

```
apt-get install Python-lxml
easy_install lxml
pip install lxml
```

另一个可以供 Beautiful Soup 使用的解析器是 html5lib，这是一款用纯 Python 语言实现的解析器。html5lib 的解析方式与浏览器相同，可以使用下面的命令来安装 html5lib：

```
apt-get install Python-html5lib
easy_install html5lib
pip install html5lib
```

2. 使用 Beautiful Soup

在下面的实例中，演示了使用 Beautiful Soup 解析网页的过程。实例文件 bs03.py 的功能是解析清华大学出版社官方网站的主页，具体实现代码如下。

源码路径：codes\1\1-2\bs03.py

```
import urllib.request
from bs4 import BeautifulSoup

url = "http://www.tup.tsinghua.edu.cn/index.html"
page = urllib.request.urlopen(url)
soup = BeautifulSoup(page,"html.parser")
print(soup)
```

执行后会显示解析后获得的网页代码，如图 1-5 所示。

```
C:\ProgramData\Anaconda3\python.exe E:/123/codes--data/1/1-2/bs03.py

<!DOCTYPE html>

<html lang="en">
<head><meta charset="utf-8"/><title>
    清华大学出版社
</title><meta content="text/html; charset=utf-8" http-equiv="Content-Type"/><meta content="IE=8" http-equiv="X-UA-Compatible"/><meta
<script src="js/xslb/jquery-1.4.2.min.js" type="text/javascript"></script>
</head>
<body>
<div class="nav">
<div class="logo">
<div class="logo1">
<ul>
<li class="logozc"><a href="member/register.aspx">注册</a></li>
<li>
<div class="logofh">
                +</div>
            <a href="member/dl.aspx">登录</a></li>
</ul>
</div>
<div class="logo2">
<div class="logo21">
<img alt="" src="img/logo.jpg"/></div>
<div class="logo22">
<div class="logo22-1 bxs">
            书问搜索</div>
<div class="logo22-2 bxs">
            站内搜索</div>
```

图 1-5　解析结果的部分解图

在下面的实例中，演示了使用 Beautiful Soup 解析并获取网页信息的过程。

实例文件 bs04.py 的功能是，解析并获取清华大学出版社官网中最新的 15 本 Python 图书信息，并打印输出这 15 本 Python 图书的基本信息。文件 bs04.py 的具体实现代码如下。

源码路径：codes\1\1-2\bs04.py

```
from urllib.request import urlopen
from bs4 import BeautifulSoup
html=urlopen("http://www.tup.tsinghua.edu.cn/booksCenter/booklist.html?keyw
ord=python&keytm=8E323D219188916A8F")
bsObj=BeautifulSoup(html,"html.parser")

nameList=bsObj.findAll("ul",{"class":"b_list"})
for name in nameList:
    print(name.get_text()))
```

通过上述代码，可以解析清华大学出版社官网中关键字为 python 的图书信息，只是抓取了第一个分页的 15 本图书的书名。执行后会输出：

```
Python 3.8 从入门到精通(视频... 王英英 9787302552116 定价:89 元 Python 机器学习及实践 梁
佩莹 9787302539735 定价: 79 元 Python 案例教程 钱毅湘、熊福松 9787302550587 定价: 39 元
```

Python程序设计 黄蔚 熊福松 9787302550235 定价：59 元青少年学 Python 编程(配套视频...龙豪杰 9787302552123 定价：59 元 Python 机器学习算法与应用 邓立国 9787302548997 定价：69 元 Python 3.8 从零开始学 刘宇宙、刘艳 9787302552147 定价：79.80 元深度学习:从 Python 到 TensorFlo... 叶虎 9787302545651 定价：69.80 元 Python 数据科学零基础一本通 洪锦魁 9787302545392 定价：129 元 Python 语言程序设计 陈振 9787302547860 定价：49 元 Python 爬虫大数据采集与挖掘-... 曾剑平 9787302540540 定价：59.80 元中学生 Python 与 micro: bit 机器... 高旸、尚凯 9787302537625 定价：49 元 Python 人工智能 刘伟善 9787302547792 定价：59.80 元 Python 数据分析与可视化-微课... 魏伟一、李晓红 9787302546665 定价：49 元 Python 程序设计 翟萍、王军锋、9787302544388 定价：44.50 元

> **注 意**
> 因为清华大学出版社官网中的数据是实时变化的，所以每次执行上述程序后的输出结果会有所不同。

1.2.3 使用 XPath 爬取网络数据

在 Python 程序中，可以使用 XPath 来解析爬虫数据。XPath 不但提供了非常简洁、明了的路径选择表达式，而且还提供了超过上百个用于处理字符串、数值、时间的匹配以及节点、序列的内置函数。

1. 安装 XPath

XPath 是一种在 XML 文档中查找信息的语言，可以遍历并提取 XML 文档中的元素和属性。可以使用以下命令安装 XPath：

```
pip install lxml
```

在 Python 程序中使用 XPath 时，需要通过以下命令导入 Xpath：

```
from lxml import etree
```

2. 使用 XPath

在下面的实例中，演示了使用 XPath 解析一段 HTML 代码的过程。实例文件 xp01.py 的具体实现代码如下。

> 源码路径：codes\1\1-2\xp01.py

```
from lxml import etree

wb_data = """
        <div>
            <ul>
                <li class="item-0"><a href="link1.html">first item</a></li>
                <li class="item-1"><a href="link2.html">second item</a></li>
                <li class="item-inactive"><a href="link3.html">third item</a></li>
                <li class="item-1"><a href="link4.html">fourth item</a></li>
                <li class="item-0"><a href="link5.html">fifth item</a>
            </ul>
        </div>
        """
```

```
html = etree.HTML(wb_data)
print(html)
result = etree.tostring(html)
print(result.decode("utf-8"))
```

执行后会输出：

```
<Element html at 0x1acbac0b708>
<html><body><div>
        <ul>
            <li class="item-0"><a href="link1.html">first item</a></li>
            <li class="item-1"><a href="link2.html">second item</a></li>
            <li class="item-inactive"><a href="link3.html">third item</a></li>
            <li class="item-1"><a href="link4.html">fourth item</a></li>
            <li class="item-0"><a href="link5.html">fifth item</a>
        </li></ul>

    </div>
        </body></html>
```

由此可见，XPath 会在解析结果中自动补全缺少的 HTML 标记元素。而在下面的实例中，演示了使用 XPath 解析 HTML 代码中的指定标签的过程。

实例文件 xp02.py 的具体实现代码如下。

源码路径：codes\1\1-2\xp02.py

```
from lxml import etree

wb_data = """
        <div>
            <ul>
                <li class="item-0"><a href="link1.html">first item</a></li>
                <li class="item-1"><a href="link2.html">second item</a></li>
                <li class="item-inactive"><a href="link3.html">third item</a></li>
                <li class="item-1"><a href="link4.html">fourth item</a></li>
                <li class="item-0"><a href="link5.html">fifth item</a>
            </ul>
        </div>
        """
html = etree.HTML(wb_data)
html_data = html.xpath('/html/body/div/ul/li/a')
print(html)
for i in html_data:
    print(i.text)
```

执行后会解析出 html/body/div/ul/li/a 标签下的内容：

```
first item
second item
third item
fourth item
fifth item
```

在下面的实例中，演示了使用 XPath 解析并获取网页数据的过程。

实例文件 xp03.py 的功能是，解析并获取清华大学出版社官方网站中"重点推荐"的作者信息，并打印输出"重点推荐"的作者信息。文件 xp03.py 的具体实现代码如下。

> 源码路径：codes\1\1-2\xp03.py

```
from lxml import etree

html =
etree.parse('http://www.tup.tsinghua.edu.cn/booksCenter/books_index.html',
etree.HTMLParser())
result = html.xpath('//div[@class="m_b_right"]/p/text()')
print(result)
```

执行上述代码会打印输出清华大学出版社官方网站中"重点推荐"的作者信息：

['菠萝，本名李治中，清华大学生物系本科，美国杜克大学癌症生物学博士，现担任美国诺华制药癌症新药开发部资深研究员，实验室负责人。爱好科普和公益事业。"健康不是闹着玩儿"公众号运营者之一，"向日葵"儿童癌症公益平台发起者。\r\n   ', '\r\n   ']

> **注 意**
> 因为清华大学出版社官网中的数据是实时变化的，所以每次执行上述程序后的输出结果会有所不同。

1.2.4 爬取体育新闻信息并保存到 XML 文件

在本节的内容中，将通过一个具体实例的实施过程，详细讲解使用 Python 爬取新闻信息并保存到 XML 文件中的方法，然后讲解了使用 Stanford CoreNLP 提取 XML 数据特征关系的过程。

1. 爬虫爬取数据

在本项目的 Scrap 目录中提供了多个爬虫文件，每个文件都可以爬取指定网页的新闻信息，并且都可以将爬取的信息保存到 XML 文件中。例如，通过文件 scrap1.py 可以抓取新浪体育某个页面中的新闻信息，并将抓取的信息保存到 XML 文件 news1.xml 中。文件 scrap1.py 的主要实现代码如下。

> 源码路径：codes\1\1-2\pythonCrawler\venv\Scrap\scrap1.py

```
doc=xml.dom.minidom.Document()
root=doc.createElement('AllNews')
doc.appendChild(root)
#用于爬取新浪体育的网页
urls2=['http://sports.example..com.cn/nba/25.shtml']

def scrap():
    for url in urls2:
        count = 0   # 用于统计总共爬取的新闻数量
        html = urlopen(url).read().decode('utf-8')
        #print(html)
        res=re.findall(r'<a href="(.*?)" target="_blank">(.+?)</a><br><span>',html)#用于爬取超链接和新闻标题
```

```
        for i in res:
            try:
                urli=i[0]
                htmli=urlopen(urli).read().decode('utf-8')
                time=re.findall(r'<span class="date">(.*?)</span>',htmli)
                resp=re.findall(r'<p>(.*?)</p>',htmli)
                #subHtml=re.findall('',htmli)
                nodeNews=doc.createElement('News')
                nodeTopic=doc.createElement('Topic')
                nodeTopic.appendChild(doc.createTextNode('sports'))
                nodeLink=doc.createElement('Link')
                nodeLink.appendChild(doc.createTextNode(str(i[0])))
                nodeTitle=doc.createElement('Title')
                nodeTitle.appendChild(doc.createTextNode(str(i[1])))
                nodeTime=doc.createElement('Time')
                nodeTime.appendChild(doc.createTextNode(str(time)))
                nodeText=doc.createElement('Text')
                nodeText.appendChild(doc.createTextNode(str(resp)))
                nodeNews.appendChild(nodeTopic)
                nodeNews.appendChild(nodeLink)
                nodeNews.appendChild(nodeTitle)
                nodeNews.appendChild(nodeTime)
                nodeNews.appendChild(nodeText)
                root.appendChild(nodeNews)
                print(i)
                print(time)
                print(resp)
                count+=1
            except:
                print(count)
                break
scrap()
fp=open('news1.xml','w', encoding="utf-8")
doc.writexml(fp, indent='', addindent='\t', newl='\n', encoding="utf-8")
```

执行后会将爬取的新浪体育的新闻信息保存到 XML 文件 news1.xml 中，如图 1-6 所示。

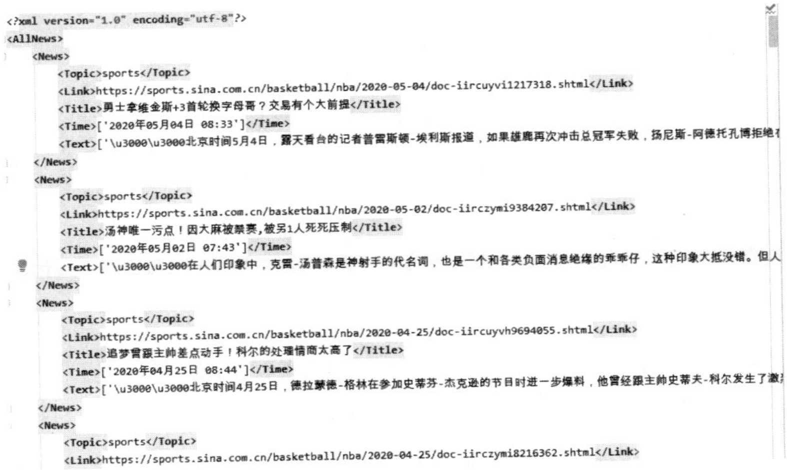

```
<?xml version="1.0" encoding="utf-8"?>
<AllNews>
	<News>
		<Topic>sports</Topic>
		<Link>https://sports.sina.com.cn/basketball/nba/2020-05-04/doc-iircuyvi1217318.shtml</Link>
		<Title>男士拿维金斯+3首轮换字母哥？交易有个大前提</Title>
		<Time>['2020年05月04日 08:33']</Time>
		<Text>['\u3000\u3000北京时间5月4日，露天看台的记者普雷斯顿-埃利斯报道，如果雄鹿再次冲击总冠军失败，扬尼斯-阿德托孔博拒绝
	</News>
	<News>
		<Topic>sports</Topic>
		<Link>https://sports.sina.com.cn/basketball/nba/2020-05-02/doc-iirczymi9384207.shtml</Link>
		<Title>汤神唯一污点！因大麻被禁赛,被另1人死死压制</Title>
		<Time>['2020年05月02日 07:43']</Time>
		<Text>['\u3000\u3000在人们印象中，克雷-汤普森是神射手的代名词，也是一个和各类负面消息绝缘的乖乖仔，这种印象大抵没错。但人
	</News>
	<News>
		<Topic>sports</Topic>
		<Link>https://sports.sina.com.cn/basketball/nba/2020-04-25/doc-iircuyvh9694055.shtml</Link>
		<Title>追梦曾跟主帅差点动手！科尔的处理情商太高了</Title>
		<Time>['2020年04月25日 08:44']</Time>
		<Text>['\u3000\u3000北京时间4月25日，德拉蒙德-格林在参加史蒂芬-杰克逊的节目时进一步爆料，他曾经跟主帅史蒂夫-科尔发生了激
	</News>
	<News>
		<Topic>sports</Topic>
		<Link>https://sports.sina.com.cn/basketball/nba/2020-04-25/doc-iirczymi8216362.shtml</Link>
```

图 1-6　文件 news1.xml

2. 使用 Stanford CoreNLP 提取 XML 数据的特征关系

Stanford CoreNLP 是由斯坦福大学开源的一套 Java NLP 工具，提供了词性标注(Part-of-speech (POS) tagger)、命名实体识别(Named Entity Recognizer，NER)、情感分析(Sentiment Analysis)等功能。Stanford CoreNLP 为 Python 提供了对应的模块，可以通过以下命令进行安装：

```
pip install stanfordcorenlp
```

因为本项目抓取到的是中文信息，所以还需要在 Stanford CoreNLP 官网下载专门处理中文的软件包，如 stanford-chinese-corenlp-2018-10-05-models.jar。

编写文件 nlpTest.py，功能是调用 Stanford CoreNLP 分析处理上面抓取到的数据文件 news1.xml，提取出数据中的人名、城市和组织等信息，主要实现代码如下。

源码路径：codes\1\1-2\pythonCrawler\venv\nlpTest.py

```python
import os
from stanfordcorenlp import StanfordCoreNLP

nlp = StanfordCoreNLP(r'H:\stanford-corenlp-full-2018-10-05', lang='zh')

for line in open(r'H:\pythonshuju\2\1-5\pythonCrawler-master\venv\Scrap\news1.xml','r', encoding='utf-8'):
    res=nlp.ner(line)
    person = ["PERSON:"]
    location=['LOCATION:']
    organization=['ORGNIZATION']
    gpe=['GRE']
    for i in range(len(res)):
        if res[i][1]=='PERSON':
            person.append(res[i][0])
    for i in range(len(res)):
        if res[i][1]=='LOCATION':
            location.append(res[i][0])
    for i in range(len(res)):
        if res[i][1]=='ORGANIZATION':
            organization.append(res[i][0])
    for i in range(len(res)):
        if res[i][1]=='GPE':
            gpe.append(res[i][0])
    print(person)
    print(location)
    print(organization)
    print(gpe)

nlp.close()
```

执行后会输出提取分析后的数据：

```
['PERSON:', '凯文', '杜兰特', '杜兰特', '金州', '杜兰特', '曼尼-迪亚兹', 'Manny', 'Diaz', '迪亚兹', '杜兰特', '杜兰特', '威廉姆森', '巴雷特和卡', '雷蒂什']
```

```
['LOCATION:']
['ORGNIZATION', 'NCAA', '球队', '迈阿密', '大学', '橄榄', '球队', '迈阿密', '大学
', '新任', '橄榄球队', '迈阿密', '先驱者', '报', '杜克大学']
#######后面省略好多信息
```

1.2.5 爬取××百科

本实例文件 baike.py 能够爬取××百科网站中的热门信息，具体功能如下。

- 抓取××百科热门段子。
- 过滤带有图片的段子。

每当按一次回车键，便会显示一个段子的发布时间、发布人、段子内容和点赞数。

🅔 源码路径：**daima\1\1-2\baike.py**

(1) 确定 URL 并爬取页面代码。

首先确定好页面的 URL 是 http://www.域名主页.com/hot/page/1，其中最后一个数字 1 代表当前的页数，可以传入不同的值来获得某一页的段子内容。首先编写代码设置要抓取的目标首页和 user_agent 值，具体实现代码如下：

```
import urllib
import urllib.request
page = 1
url = 'http://www.域名主页.com/hot/page/' + str(page)
user_agent = 'Mozilla/4.0 (compatible; MSIE 5.5; Windows NT)'
headers = { 'User-Agent' : user_agent }
try:
    request = urllib.request.Request(url,headers = headers)
    response = urllib.request.urlopen(request)
    print (response.read())
except (urllib.request.URLError, e):
    if hasattr(e,"code"):
        print (e.code)
    if hasattr(e,"reason"):
        print e.reason
```

执行后会打印输出第一页的 HTML 代码。

(2) 提取某一页的所有段子。

在获取 HTML 代码后，接下来开始获取某一页的所有段子。首先审查元素看一下，按浏览器的 F12 键，截图如图 1-7 所示。

```
<div class="articleGender manIcon">28</div>
</div>


<a href="/article/118348358" target="_blank" class='contentHerf' >
<div class="content">

<span>历史课上，女老师的鞋跟突然断了。她感慨道："我这鞋都穿五年了。"这时，角落里有人低声说道："不愧是历史老师！"</span>


</div>
</a>
```

图 1-7 http://www.域名主页.com/hot/的源码

由此可见，每个段子都是被<div class="articleGender manIcon">...</div>包含的内容。如果想获取页面中的发布人、发布日期、段子内容及点赞个数，需要注意有些段子是带有图片的。因为不能在控制台中显示图片信息，所以需要删除带有图片的段子，确保保存只含文本的段子。为了实现这一功能，使用正则表达式方法 re.findall( ) 寻找所有匹配的内容。编写的正则表达式匹配语句如下：

```
    pattern = re.compile(
        '<div.*?author clearfix">.*?<h2>(.*?)</h2>.*?<div.*?content".*?<span>(.*?)</span>.*?</a>(.*?)<div class= "stats".*?number">(.*?)</i>', re.S)
```

上述正则表达式是整个程序的核心，本实例的实现文件是 baike.py，具体实现代码如下：

```python
import urllib.request
import re
class Qiubai:

    # 初始化，定义一些变量
    def __init__(self):
        # 初始页面为1
        self.pageIndex = 1
        # 定义UA
        self.user_agent = 'Mozilla/5.0 (Windows NT 6.1; WOW64) AppleWebKit/537.36 (KHTML, like Gecko) Chrome/55.0.2883.75 Safari/537.36'
        # 定义 headers
        self.headers = {'User-Agent': self.user_agent}
        # 存放段子的变量，每个元素是每一页的段子
        self.stories = []
        # 程序是否继续运行
        self.enable = False
    def getPage(self, pageIndex):
        """
        传入某一页面索引后的页面代码
        """
        try:
            url = 'http://www.域名主页.com/hot/page/' + str(pageIndex)
            # 构建 request
            request = urllib.request.Request(url, headers=self.headers)
            # 利用 urlopen 获取页面代码
            response = urllib.request.urlopen(request)
            # 页面转为utf-8 编码
            pageCode = response.read().decode("utf8")
            return pageCode
        # 捕获错误原因
        except (urllib.request.URLError, e):
            if hasattr(e, "reason"):
                print (u"连接糗事百科失败，错误原因", e.reason)
                return None
    def getPageItems(self, pageIndex):
        """
        传入某一页代码，返回本页不带图的段子列表
        """
        # 获取页面代码
```

```python
            pageCode = self.getPage(pageIndex)
            # 如果获取失败，返回 None
            if not pageCode:
                print ("页面加载失败...")
                return None
            # 匹配模式
            pattern = re.compile(
                '<div.*?author clearfix">.*?<h2>(.*?)</h2>.*?<div.*?content".*?<span>(.*?)</span>.*?</a>(.*?)<div class= "stats".*?number">(.*?)</i>', re.S)
            # findall 匹配整个页面内容,items 匹配结果
            items = re.findall(pattern, pageCode)
            # 存储整页的段子
            pageStories = []
            # 遍历正则表达式匹配的结果, 0 name, 1 content, 2 img, 3 votes
            for item in items:
                # 是否含有图片
                haveImg = re.search("img", item[2])
                # 不含，加入 pageStories
                if not haveImg:
                    # 替换 content 中的<br/>标签为\n
                    replaceBR = re.compile('<br/>')
                    text = re.sub(replaceBR, "\n", item[1])
                    # 在 pageStories 中存储: 名字、内容、赞数
                    pageStories.append(
                        [item[0].strip(), text.strip(), item[3].strip()])
            return pageStories
        def loadPage(self):
            """
            加载并提取页面的内容，加入列表
            """
            # 如未看页数少于 2, 则加载并抓取新一页补充
            if self.enable is True:
                if len(self.stories) < 2:
                    pageStories = self.getPageItems(self.pageIndex)
                    if pageStories:
                        # 添加到 self.stories 列表中
                        self.stories.append(pageStories)
                        # 实际访问的页码+1
                        self.pageIndex += 1
        def getOneStory(self, pageStories, page):
            """
            调用该方法，回车输出段子，按 q 结束程序的运行
            """
            # 循环访问一页的段子
            for story in pageStories:
                # 等待用户输入，回车输出段子, q 退出
                shuru = input()
                self.loadPage()
                # 如果用户输入 q 退出
                if shuru == "q":
                    # 停止程序运行, start()中 while 判定
                    self.enable = False
                    return
                # 打印 story:0 name, 1 content, 2 votes
```

```python
        print (u"第%d 页\t 发布人:%s\t\3:%s\n%s" % (page, story[0], story[2], story[1]))
    def start(self):
        """
        开始方法
        """
        print (u"正在读取糗事百科，回车查看新段子，q 退出")
        # 程序运行变量 True
        self.enable = True
        # 加载一页内容
        self.loadPage()
        # 局部变量，控制当前读到了第几页
        nowPage = 0
        # 直到用户输入 q, self.enable 为 False
        while self.enable:
            if len(self.stories) > 0:
                # 吐出一页段子
                pageStories = self.stories.pop(0)
                # 用于打印当前页面，当前页数+1
                nowPage += 1
                # 输出这一页段子
                self.getOneStory(pageStories, nowPage)
if __name__ == '__main__':
    qiubaiSpider = Qiubai()
    qiubaiSpider.start()
```

执行效果如图 1-8 所示，每次按回车键就会显示下一条热门糗事信息。

图 1-8　执行效果

1.3　使用专业爬虫库 Scrapy

因为爬虫应用程序的需求日益高涨，所以在市面上诞生了很多第三方开元爬虫框架，其中 Scrapy 是一个为了爬取网站数据、提取结构性数据而编写的专业框架。Scrapy 框架的用途十分广泛，可以用于数据挖掘、数据监测和自动化测试等工作。本节将简要讲解爬虫框架 Scrapy 的基本用法。

扫码观看本节视频讲解

1.3.1 Scrapy 框架基础

框架 Scrapy 使用了 Twisted 异步网络库来处理网络通信，其整体架构大致如图 1-9 所示。

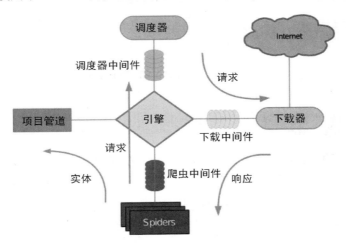

图 1-9　框架 Scrapy 的架构

在 Scrapy 框架中，主要包括以下组件。
- Scrapy 引擎：用来处理整个系统的数据流，会触发的框架核心事务。
- 调度器：用来获取 Scrapy 发送过来的请求，然后将请求传入队列，并在引擎再次请求时返回。调度器的功能是设置下一个要抓取哪一个网址，并且还可以删除重复的网址。
- 下载器：建立在高效的异步模型 Twisted 之上，功能是下载目标网址中的网页内容，并将网页内容返回给 Scrapy。
- 爬虫(Spiders)：功能是从特定的网页中提取指定的信息，这些信息在爬虫领域中称为实体。
- 项目管道：功能是处理从网页中提取的爬虫实体。当使用爬虫解析一个页面后，会将实体发送到项目管道中进行处理，然后验证实体的有效性，并将不需要的信息删除。
- 下载器中间件：此模块位于 Scrapy 引擎和下载器之间，为 Scrapy 引擎与下载器之间的请求及响应建立桥梁。
- 爬虫中间件：此模块在 Scrapy 引擎和 Spiders 之间，功能是处理爬虫的响应输入和请求输出。
- 调度中间件：在 Scrapy 引擎和调度器之间，表示从 Scrapy 引擎发送到调度的请求和响应。

在使用 Scrapy 框架后，下面是大多数爬虫程序的运行流程。
- Scrapy 引擎从调度器中取出一个 URL 链接,这个链接将会作为接下来要抓取目标。

- Scrapy 引擎将目标 URL 封装成一个 Request 并传递给下载器，下载器在下载 URL 资源后，将资源封装成应答包。
- 使用爬虫解析应答包，如果解析出实体，则将结果交给项目管道作进一步处理。如果解析出的是 URL 链接，则把 URL 交给调度器等待下一步的抓取操作。

1.3.2 搭建 Scrapy 环境

在本地计算机安装 Python 后，可以使用 pip 命令或 easy_install 命令安装 Scrapy，具体命令格式如下：

```
pip install scrapy
easy_install scrapy
```

另外，还需要确保已经安装了 win32api 模块，在安装此模块时必须安装和本地 Python 版本相对应的版本和位数(32 位或 64 位)。读者可以登录：http://www.lfd.uci.edu/~gohlke/pythonlibs/找到需要的版本，如图 1-10 所示。

PyWin32: extensions for Windows.
To install pywin32 system files, run `*python.exe Scripts/pywin32_postinstall.py -install*` from an elevated command prompt.
pywin32-227-cp38-cp38-win_amd64.whl
pywin32-227-cp38-cp38-win32.whl
pywin32-227-cp37-cp37m-win_amd64.whl
pywin32-227-cp37-cp37m-win32.whl
pywin32-227-cp36-cp36m-win_amd64.whl
pywin32-227-cp36-cp36m-win32.whl
pywin32-227-cp35-cp35m-win_amd64.whl
pywin32-227-cp35-cp35m-win32.whl
pywin32-227-cp27-cp27m-win_amd64.whl
pywin32-227-cp27-cp27m-win32.whl
pywin32-224-cp34-cp34m-win_amd64.whl
pywin32-224-cp34-cp34m-win32.whl
pypiwin32-224+dummy-py2.py3-none-any.whl

图 1-10 下载 win32api 模块

下载后将得到一个.whl 格式的文件，定位到此文件的目录，然后通过以下命令即可安装 win32api 模块：

```
python -m pip install --user ".whl"格式文件的全名
```

注 意

如果出现"ImportError: DLL load failed: 找不到指定的模块。"错误，需要将 Python\Python37\Lib\site-packages\win32 目录中的以下文件保存到本地系统盘中的 Windows\System32 目录下：

- pythoncom37.dll
- pywintypes37.dll

1.3.3 创建第一个 Scrapy 项目

下面的实例代码，演示了创建第一个 Scrapy 项目的过程。

源码路径：daima\1\1-3\tutorial

(1) 创建项目。

在开始爬取数据之前，必须先创建一个新的 Scrapy 项目。进入准备存储代码的目录中，然后运行以下命令：

```
scrapy startproject tutorial
```

上述命令的功能是创建一个包含下列内容的 tutorial 目录：

```
tutorial/
    scrapy.cfg
    tutorial/
        __init__.py
        items.py
        pipelines.py
        settings.py
        spiders/
            __init__.py
            ...
```

对上述文件的具体说明如下。

- scrapy.cfg：项目的配置文件。
- tutorial/：该项目的 python 模块。之后可以在此加入代码。
- tutorial/items.py：项目中的 item 文件。
- tutorial/pipelines.py：项目中的 pipelines 文件。
- tutorial/settings.py：项目的设置文件。
- tutorial/spiders/：放置 spider 代码的目录。

(2) 定义 Item。

Item 是保存爬取到的数据的容器，其使用方法和 Python 中的字典类似，并且提供了额外保护机制来避免拼写错误导致的未定义字段错误。可以通过创建 scrapy.Item 类，并且定义类型为 scrapy.Field 的类属性来定义 Item。

首先根据需要从 dmoz.org 获取到的数据对 item 进行建模。需要从 dmoz 中获取名字：URL 以及网站的描述。对此，在 item 中定义相应的字段。编辑 tutorial 目录中的文件 items.py，具体实现代码如下：

```
import scrapy
class DmozItem(scrapy.Item):
    title = scrapy.Field()
    link = scrapy.Field()
    desc = scrapy.Field()
```

通过定义 item，可以很方便地使用 Scrapy 中的其他方法。而这些方法需要知道我们的 item 的定义。

(3) 编写第一个爬虫(Spider)。

Spider 是用户编写用于从单个网站(或者一些网站)爬取数据的类，其中包含一个用于下载的初始 URL，如何跟进网页中的链接以及如何分析页面中的内容，提取生成 item 的方法。为了创建 Spider，必须继承类 scrapy.Spider，且定义以下 3 个属性。

- name：用于区别 Spider。该名字必须是唯一的，不可以为不同的 Spider 设定相同的名字。
- start_urls：包含了 Spider 在启动时进行爬取的 URL 列表。因此，第一个被获取到的页面将是其中之一。后续的 URL 则从初始的 URL 获取到的数据中提取。
- parse( )：是 Spider 的一个方法。被调用时，每个初始 URL 完成下载后生成的 Response 对象将会作为唯一的参数传递给该函数。该方法负责解析返回的数据 (response data)、提取数据(生成 item)以及生成需要进一步处理的 URL 的 Request 对象。

下面是编写的第一个 Spider 代码，保存在 tutorial/spiders 目录下的文件 dmoz_spider.py 中，具体实现代码如下：

```
import scrapy
class DmozSpider(scrapy.Spider):
    name = "dmoz"
    allowed_domains = ["dmoz.org"]
    start_urls = [
        "http://www.dmoz.org/Computers/Programming/Languages/Python/Books/",
        "http://www.dmoz.org/Computers/Programming/Languages/Python/Resources/"
    ]
    def parse(self, response):
        filename = response.url.split("/")[-2]
        with open(filename, 'wb') as f:
            f.write(response.body)
```

(4) 爬取。

进入项目的根目录，执行下列命令启动 Spider：

```
scrapy crawl dmoz
```

crawl dmoz 是负责启动用于爬取 dmoz.org 网站的 Spider，之后会得到以下输出：

```
2019-01-23 18:13:07-0400 [scrapy] INFO: Scrapy started (bot: tutorial)
2019-01-23 18:13:07-0400 [scrapy] INFO: Optional features available: ...
2019-01-23 18:13:07-0400 [scrapy] INFO: Overridden settings: {}
2019-01-23 18:13:07-0400 [scrapy] INFO: Enabled extensions: ...
2019-01-23 18:13:07-0400 [scrapy] INFO: Enabled downloader middlewares: ...
2019-01-23 18:13:07-0400 [scrapy] INFO: Enabled spider middlewares: ...
2019-01-23 18:13:07-0400 [scrapy] INFO: Enabled item pipelines: ...
2019-01-23 18:13:07-0400 [dmoz] INFO: Spider opened
2019-01-23 18:13:08-0400 [dmoz] DEBUG: Crawled (200) <GET http://www.dmoz.org/Computers/Programming/Languages/Python/Resources/> (referer: None)
2019-01-23 18:13:09-0400 [dmoz] DEBUG: Crawled (200) <GET http://www.dmoz.org/Computers/Programming/Languages/Python/Books/> (referer: None)
2019-01-23 18:13:09-0400 [dmoz] INFO: Closing spider (finished)
```

查看包含 dmoz 的输出，可以看到在输出的 log 中包含定义在 start_urls 的初始 URL，并且与 Spider 中是一一对应的。在 log 中可以看到其没有指向其他页面((referer:None))。此外，创建了两个包含 URL 所对应的内容的文件，即 Book 和 Resources。

由此可见，Scrapy 为 Spider 的 start_urls 属性中的每个 URL 创建了 scrapy.Request 对象，并将 parse 方法作为回调函数(callback)赋值给 Request。Request 对象经过调度，执行生成 scrapy.http.Response 对象并送回给 spider parse( )方法。

（5）提取 Item。

有很多种从网页中提取数据的方法，Scrapy 使用一种基于 XPath 和 CSS 表达式机制，即 Scrapy Selectors。关于 Selector 和其他提取机制的信息，建议读者参考 Selector 的官方文档。下面给出 XPath 表达式及其含义。

- /html/head/title：选择 HTML 文档中<head>标签内的<title>元素。
- /html/head/title/text( )：选择上面提到的<title>元素的文字。
- //td：选择所有的<td>元素。
- //div[@class="mine"]：选择所有具有 class="mine"属性的 div 元素。

上边仅列出了几个简单的 XPath 例子，XPath 实际上要比这远远强大得多。为了配合 XPath，Scrapy 除了提供 Selector 外，还提供了其他方法来避免每次从 Response 中提取数据时生成 Selector 的麻烦。

在 Selector 中有以下几个最基本的方法。

- xpath( )：用于选取指定的标签内容，如下面的代码表示选取所有的 book 标签：

```
selector.xpath('//book')
```

- css( )：传入 CSS 表达式，用于选取指定的 CSS 标签内容。
- extract( )：返回选中内容的 Unicode 字符串，返回结果是列表。
- re( )：根据传入的正则表达式提取数据，返回 Unicode 字符串格式的列表。
- re_first( )：返回 SelectorList 对象中的第一个 Selector 对象调用的 re 方法。

接下来使用内置的 Scrapy shell，首先进入本实例项目的根目录，然后执行以下命令来启动 shell：

```
scrapy shell "http://www.dmoz.org/Computers/Programming/Languages/Python/Books/"
```

此时 shell 将会输出类似以下内容：

```
[ ... Scrapy log here ... ]
2019-01-23 17:11:41-0400 [default] DEBUG: Crawled (200) <GET http://www.dmoz.org/Computers/Programming/Languages/Python/Books/> (referer: None)
[s] Available Scrapy objects:
[s]   crawler    <scrapy.crawler.Crawler object at 0x3636b50>
[s]   item       {}
[s]   request    <GET http://www.dmoz.org/Computers/Programming/Languages/Python/Books/>
[s]   response   <200 http://www.dmoz.org/Computers/Programming/Languages/Python/Books/>
[s]   settings   <scrapy.settings.Settings object at 0x3fadc50>
[s]   spider     <Spider 'default' at 0x3cebf50>
[s] Useful shortcuts:
[s]   shelp()           Shell help (print this help)
[s]   fetch(req_or_url) Fetch request (or URL) and update local objects
```

```
[s]   view(response)    View response in a browser
In [1]:
```

当载入 shell 后会得到一个包含 response 数据的本地 response 变量。输入 response.body 命令后会输出 response 的包体，输入 response.headers 后可以看到 response 的包头。更为重要的是，当输入 response.selector 时，将获取到一个可以用于查询返回数据的 selector(选择器)以及映射到 response.selector.xpath()、response.selector.css() 的快捷方法(shortcut):response.xpath()和 response.css()。同时，shell 根据 response 提前初始化了变量 sel。该 selector 根据 response 的类型自动选择最合适的分析规则(XML vs HTML)。

(6) 提取数据。

接下来尝试从这些页面中提取一些有用的数据，可以在终端中输入 response.body 来观察 HTML 源码并确定合适的 XPath 表达式。但是这个任务非常无聊且不易，可以考虑使用 Firefox 的 Firebug 扩展来简化工作。

在查看网页的源码后，会发现网站的信息包含在第二个<ul>元素中。可以通过下面的代码选择该页面中网站列表里的所有<li>元素：

```
response.xpath('//ul/li')
```

通过以下命令获取对网站的描述：

```
response.xpath('//ul/li/text()').extract()
```

通过以下命令获取网站的标题：

```
response.xpath('//ul/li/a/text()').extract()
```

1.3.4 爬取某电影网的热门电影信息

本实例的功能是，使用 Scrapy 爬虫抓取某电影网中热门电影信息的过程。

> 源码路径：daima\1\1-3\douban

(1) 在具体爬取数据之前，必须先创建一个新的 Scrapy 项目。首先进入准备保存项目代码的目录中，然后运行以下命令：

```
scrapy startproject scrapydouban
```

(2) 编写文件 moviedouban.py 设置要爬取的 URL 范围和过滤规则，主要实现代码如下：

```
class MoviedoubanSpider(CrawlSpider):
    name = "moviedouban"
    allowed_domains = ["movie.域名主页.com"]
    start_urls = ["https://movie.域名主页.com/"]

    rules = (
        Rule(LinkExtractor(allow=r"/subject/\d+/($|\?|\w+)"),
            callback="parse_movie", follow=True),
    )

    def __init__(self):
```

```
        self.page_number = 1

    def parse_movie(self, response):
        print("RESPONSE: {}".format(response))
```

(3) 编写执行脚本文件 pyrequests_douban.py，功能是编写功能函数获取热门电影的详细信息，包括电影名、URL 链接地址、导演信息、主演信息等。文件 pyrequests_douban.py 的主要实现代码如下：

```
def get_celebrity(url):
    print("Sending request to {}".format(url))
    celebrity = {}
    res = get_request(url)
    if res.status_code == 200:
        html = BeautifulSoup(res.content, "html.parser")
        celebrity['avatars'] = {}
        if html.find("div", {'class': "pic"}):
            for s in ["small", "medium", "large"]:
                celebrity['avatars'][s] = [dt['src'].strip() for dt in html.find("div", {'class': "pic"}).findAll("img", src=re.compile(r'{}'.format(s)))]
    return celebrity


def get_initial_release(html):
    return html.find("div", {'id': "info"}).find("span", {'property': "v:initialReleaseDate"}).string


def get_photos(html, url):
    photos = []
    if html.find("a", {"href": url}):
        print("Sending request for {} to get all photos".format(url))
        res = get_request(url)
        if res.status_code == 200:
            html = BeautifulSoup(res.content, 'html.parser')
            if html.find("div", {'class': "article"}):
                all_photos_html = html.find("div", {'class': "article"}).findAll("div", {'class': "mod"})[0]
                for li in all_photos_html.findAll("li")[:10]:
                    photo = {}
                    photo['url'] = li.find("a")['href']
                    photo['photo'] = li.find("img")['src']
                    photos.append(photo)
    return photos


def get_request(url, s=0):
    time.sleep(s)
    return requests.get(url)


def get_runtime(html):
```

```python
    return html.find("div", {'id': "info"}).find("span", {'property':
"v:runtime"}).string


def get_screenwriters(html):
    directors = []
    if html.find("div", {'class': "info"}):
        for d in html.find("div", {'class': "info"}).findChildren()[2].findAll("a"):
            director = {}
            director['id'] = d['href'].split('/')[2]
            director['name'] = d.string
            director['alt'] = "{}{}".format(domain, d['href'])
            director['avatars'] = get_celebrity(director['alt'])
            directors.append(director)
    return directors

def get_starring(html):
    starrings = []
    if html.find("span", {'class': "actor"}):
        for star in html.find("span", {'class': "actor"}).find("span",
{'class': "attrs"}).findAll("a"):
            starring = {}
            starring['id'] = star['href'].split('/')[2]
            starring['name'] = star.string.strip()
            starring['alt'] = "{}{}".format(domain, star['href'])
            starring['avatars'] = get_celebrity(starring['alt'])
            starrings.append(starring)
    return starrings


def get_trailer(html, url):
    trailers = []
    if html.find("a", {"href": url}):
        print("Sending request for {} to get trailers".format(url))
        res = get_request(url)
        if res.status_code == 200:
            html = BeautifulSoup(res.content, 'html.parser')
            if html.find("div", {'class': "article"}):
                for d in html.find("div", {'class': "article"}).findAll("div",
{'class': "mod"}):
                    trailer = {}
                    if "预告片" in d.find("h2").string:
                        trailer['url'] = d.find("a", {'class':
"pr-video"})['href']
                        trailer['view_img'] = d.find("a", {'class':
"pr-video"}).find("img")['src']
                        trailer['duration'] = d.find("a", {'class':
"pr-video"}).find("em").string
                        trailer['title'] =
d.find("p").find("a").string.strip()
                        trailer['date'] = d.find("p", {'class':
"trail-meta"}).find("span").string.strip()
                        trailer['responses_url'] = d.find("p", {'class':
"trail-meta"}).find("a")['href']
                        trailers.append(trailer)
```

```python
        return trailers


def get_types(html):
    return ", ".join([sp.string for sp in html.find("div", {"id": "info"}).findAll("span", {'property': "v:genre"})])


def manual_scrape():
    print("Sending request to {}".format(url.format(page_limit, page_start)))
    response = get_request(url.format(page_limit, page_start), 5)
    if response.status_code == 200:
        subjects = json.loads(response.text)['subjects']
        for subject in subjects:
            subject['api_return'] = {}
            print("Sending request to {}".format(public_api_url.format(subject['id'])))
            api_response = get_request(public_api_url.format(subject['id']))
            if api_response.status_code == 200:
                subject['api_return'] = api_response.json()
            subject['url_content'] = {}
            ## Call Happy API to post data
            print("Sending request to {}".format(subject['url']))
            content_response = get_request(subject['url'], 5)
            if content_response.status_code == 200:
                content = BeautifulSoup(content_response.content, 'html.parser')
                subject['url_content']['screenwriters'] = get_screenwriters(content)
                subject['url_content']['starring'] = get_starring(content)
                subject['url_content']['types'] = get_types(content)
                subject['url_content']['initial_release'] = get_initial_release(content)
                subject['url_content']['runtime'] = get_runtime(content)
                subject['url_content']['trailer'] = get_trailer(content, "{}trailer".format(subject['url']))
                subject['url_content']['photos'] = get_photos(content, "{}all_photos".format(subject['url']))
            pprint.pprint(subject)


if __name__ == "__main__":
    global url, page_limit, page_start, domain, public_api_url

    domain = "https://movie.域名主页.com"
    public_api_url = "https://api.域名主页.com/v2/movie/subject/{}"
    url = "https://movie.域名主页.com/j/search_subjects?type=movie&tag=%E7%83%AD%E9%97%A8&sort=recommend&page_limit={}&page_start={}"
    page_limit = 20
    page_start = 0

    manual_scrape()
```

执行后会输出显示爬取到的热门电影信息,如图 1-11 所示。

```
Sending request to https://movie.douban.com/celebrity/1373451/
Sending request for https://movie.douban.com/subject/30377703/all_photos to get all photos
{'api_return': {'aka': ['中国版《完美陌生人》','手机狂响','Kill Mobile'],
                'alt': 'https://movie.douban.com/subject/30377703/',
                'casts': [{'alt': 'https://movie.douban.com/celebrity/1009179/',
                           'avatars': {'large': 'https://img3.doubanio.com/view/celebrity/s_ratio_celebrity/public/p1542346320.44.jpg',
                                       'medium': 'https://img3.doubanio.com/view/celebrity/s_ratio_celebrity/public/p1542346320.44.jpg',
                                       'small': 'https://img3.doubanio.com/view/celebrity/s_ratio_celebrity/public/p1542346320.44.jpg'},
                           'id': '1009179',
                           'name': '佟大为'},
                          {'alt': 'https://movie.douban.com/celebrity/1319032/',
                           'avatars': {'large': 'https://img3.doubanio.com/view/celebrity/s_ratio_celebrity/public/p1444800807.11.jpg',
                                       'medium': 'https://img3.doubanio.com/view/celebrity/s_ratio_celebrity/public/p1444800807.11.jpg',
                                       'small': 'https://img3.doubanio.com/view/celebrity/s_ratio_celebrity/public/p1444800807.11.jpg'},
                           'id': '1319032',
                           'name': '马丽'},
                          {'alt': 'https://movie.douban.com/celebrity/1000145/',
                           'avatars': {'large': 'https://img3.doubanio.com/view/celebrity/s_ratio_celebrity/public/p2520.jpg',
                                       'medium': 'https://img3.doubanio.com/view/celebrity/s_ratio_celebrity/public/p2520.jpg',
                                       'small': 'https://img3.doubanio.com/view/celebrity/s_ratio_celebrity/public/p2520.jpg'},
                           'id': '1000145',
                           'name': '霍思燕'},
                          {'alt': 'https://movie.douban.com/celebrity/1316368/',
                           'avatars': {'large': 'https://img3.doubanio.com/view/celebrity/s_ratio_celebrity/public/p1473410979.5.jpg',
                                       'medium': 'https://img3.doubanio.com/view/celebrity/s_ratio_celebrity/public/p1473410979.5.jpg',
                                       'small': 'https://img3.doubanio.com/view/celebrity/s_ratio_celebrity/public/p1473410979.5.jpg'},
```

图 1-11　爬取到的热门电影信息

1.3.5　爬取某网站中的照片并保存到本地

本实例的功能是使用 Scrapy 爬虫抓取某网站中的照片信息，并将抓取到的照片保存到本地硬盘中。编写文件 art.py 设置要爬取的 URL 范围和抓取的内容元素，主要实现代码如下。

 源码路径：daima\1\1-3\ScrapyBeauty

```python
import scrapy
from scrapy import Request
import json

class ImagesSpider(scrapy.Spider):
    BASE_URL='http://域名主
页.so.com/zj?ch=beauty&sn=%s&listtype=new&temp=1'#注意这里%S 改过了
    #BASE_URL='http:// 域名主
页.so.com/j?q=%E9%BB%91%E4%B8%9D&src=srp&correct=%E9%BB%91%E4%B8%9D&pn=60&c
h=&sn=%s&sid=fc52f43bfb771f78907396c3167f10ad&ran=0&ras=0&cn=0&gn=10&kn=50'
    start_index=0

    MAX_DOWNLOAD_NUM=1000

    name="images"
    start_urls=[BASE_URL %0]

    def parse(self,response):
        infos=json.loads(response.body.decode('utf-8'))
        for info in infos['list']:
            yield {'image_urls':[info['qhimg_url']]}

        self.start_index+=30
        if infos['count']>0 and self.start_index<self.MAX_DOWNLOAD_NUM:
            yield Request(self.BASE_URL % self.start_index)
```

执行后会显示爬取目标网站图片的过程，如图 1-12 所示。

```
2019-02-11 15:31:23 [scrapy.core.engine] DEBUG: Crawled (200) <GET http://p3.so.qhmsg.com/t0154687f04bb6e87c8.jpg> (referer: None)
2019-02-11 15:31:23 [scrapy.pipelines.files] DEBUG: File (downloaded): Downloaded file from <GET http://p3.so.qhmsg.com/t0154687f04bb6e87c8.jpg> referred in <None>
2019-02-11 15:31:23 [scrapy.pipelines.files] DEBUG: File (downloaded): Downloaded file from <GET http://p4.so.qhmsg.com/t013862476b8fbf7969.jpg> referred in <None>
2019-02-11 15:31:23 [scrapy.pipelines.files] DEBUG: File (downloaded): Downloaded file from <GET http://p4.so.qhmsg.com/t01c382465efd9da959.jpg> referred in <None>
2019-02-11 15:31:23 [scrapy.core.scraper] DEBUG: Scraped from <200 http://image.so.com/zj?ch=beauty&sn=0&listtype=new&temp=1>
{'image_urls': ['http://p3.so.qhmsg.com/t0154687f04bb6e87c8.jpg'], 'images': [{'url': 'http://p3.so.qhmsg.com/t0154687f04bb6e87c8.jpg', 'path':
'full/bf820e4ed8c044f94431988051e3402f721f431f.jpg', 'checksum': 'b10e0a58b31b053eda738597b0c46722'}]}
2019-02-11 15:31:23 [scrapy.core.scraper] DEBUG: Scraped from <200 http://image.so.com/zj?ch=beauty&sn=0&listtype=new&temp=1>
{'image_urls': ['http://p4.so.qhmsg.com/t013862476b8fbf7969.jpg'], 'images': [{'url': 'http://p4.so.qhmsg.com/t013862476b8fbf7969.jpg', 'path':
'full/209d862baa417e36c248c38e39fd503d7d550eef.jpg', 'checksum': 'df568ec4c2ec907a40851fc6de25768f'}]}
2019-02-11 15:31:23 [scrapy.core.scraper] DEBUG: Scraped from <200 http://image.so.com/zj?ch=beauty&sn=0&listtype=new&temp=1>
{'image_urls': ['http://p4.so.qhmsg.com/t01c382465efd9da959.jpg'], 'images': [{'url': 'http://p4.so.qhmsg.com/t01c382465efd9da959.jpg', 'path':
'full/fe3ab046bd19a05aa8df53d7d53fe15674187000.jpg', 'checksum': 'b92548a0ed535a424124041590f712c9'}]}
2019-02-11 15:31:23 [scrapy.core.engine] DEBUG: Crawled (200) <GET http://p0.so.qhmsg.com/t017abc1de09c20ca55.jpg> (referer: None)
2019-02-11 15:31:23 [scrapy.pipelines.files] DEBUG: File (downloaded): Downloaded file from <GET http://p0.so.qhmsg.com/t017abc1de09c20ca55.jpg> referred in <None>
2019-02-11 15:31:24 [scrapy.core.scraper] DEBUG: Scraped from <200 http://image.so.com/zj?ch=beauty&sn=0&listtype=new&temp=1>
{'image_urls': ['http://p0.so.qhmsg.com/t017abc1de09c20ca55.jpg'], 'images': [{'url': 'http://p0.so.qhmsg.com/t017abc1de09c20ca55.jpg', 'path':
'full/a99c7328f98d78764752ead804ef2160bbe97ab.jpg', 'checksum': '27e9e9a362df067d33e7340571a737ee'}]}
2019-02-11 15:31:24 [scrapy.core.engine] DEBUG: Crawled (200) <GET http://p0.so.qhimgs1.com/t010165ed1120ab3834.jpg> (referer: None)
2019-02-11 15:31:24 [scrapy.pipelines.files] DEBUG: File (downloaded): Downloaded file from <GET http://p0.so.qhimgs1.com/t010165ed1120ab3834.jpg> referred in
```

图 1-12　抓取过程

并将抓取到的照片保存到本地文件夹 download_images 中，如图 1-13 所示。

图 1-13　在本地硬盘保存抓取到的照片

1.3.6　爬取某网站中的主播照片并保存到本地

本实例的功能是使用 Scrapy 抓取某网站中的主播照片，并将抓取到的主播照片保存到本地硬盘中。编写文件 douyu.py 设置要抓取的 URL 范围和抓取的内容元素，设置要抓取的 Item 是主播昵称和主播照片。文件 douyu.py 的主要实现代码如下。

源码路径：daima\1\1-3\douyu

```python
class DouyuSpider(scrapy.Spider):
    name = "douyu"
    allowed_domains = ["douyucdn.cn"]
```

```python
baseUrl = "http://capi.域名主页.cn/api/v1/getVerticalRoom?limit=208&offset="
offset = 0
start_urls = [baseUrl + str(offset),]

def parse(self, response):
    data_list = json.loads(response.body)['data']
    #在 data_list 为空的时候 ,return:关闭程序
    if len(data_list) == 0:
        return
    for data in data_list:
        item = DouyuItem()
        item['nickname'] = data["nickname"]
        item['vertical_src'] = data["vertical_src"]
        yield item
    #offset 递增 然后调用回调函数 parse()
    self.offset += 20
    scrapy.Request(self.baseUrl+str(self.offset),callback = self.parse)
```

执行后会将爬取到的主播照片保存到本地文件夹中，如图 1-14 所示。

图 1-14　抓取到的主播照片

第 2 章

使用数据库保存数据

　　数据持久化是指永久地将 Python 数据存储在磁盘上。数据持久化技术是实现动态软件项目的必需手段，在软件项目中通过数据持久化可以存储海量的数据。因为软件显示的内容是从数据库中读取的，所以开发者可以通过修改数据库内容而实现动态交互功能。在 Python 软件开发应用中，数据库在实现过程中起到中间媒介的作用。在本章的内容中，将向读者介绍 Python 数据库开发方面的核心知识，为读者步入本书后面知识的学习打下基础。

2.1 操作 SQLite 3 数据库

从 Python 3.x 版本开始,在标准库中已经内置了 sqlite3 模块,可以支持 SQLite 3 数据库的访问和相关的数据库操作。在需要操作 SQLite 3 数据库数据时,只需在程序中导入 sqlite3 模块即可。

2.1.1 sqlite3 模块介绍

扫码观看本节视频讲解

通过使用 sqlite3 模块,可以满足开发者在 Python 程序中使用 SQLite 数据库的需求。在 sqlite3 模块中包含以下常量成员。

- ▶ sqlite3.version:该 sqlite3 模块的字符串形式的版本号,这不是 SQLite 数据库的版本号。
- ▶ sqlite3.version_info:该 sqlite3 模块的整数元组形式的版本号,这不是 SQLite 数据库的版本号。
- ▶ sqlite3.sqlite_version:运行时 SQLite 数据库的版本号,是一个字符串形式。
- ▶ sqlite3.sqlite_version_info:运行时 SQLite 数据库的版本号,是一个整数元组形式。
- ▶ sqlite3.PARSE_DECLTYPES:该常量用于 connect( )函数中的 detect_types 参数,设置它使得 sqlite3 模块解析每个返回列的声明类型。将解析出声明类型的第一个单词。比如:"integer primary key"将解析出"integer",而"number(10)"将解析出"number"。
- ▶ sqlite3.PARSE_COLNAMES:该常量用于 connect( )函数中的 detect_types 参数,设置它使得 SQLite 接口解析每个返回列的列名。ga 将查找[mytype]形式的字符串,然后决定'mytype'是列的类型。将会尝试在转换器字典中找到对应为'mytype'的转换器,然后将转换器函数应用于返回值。在 Cursor.description 中找到的列名只是列名的第一个单词,如果在 SQL 中有类似'as "x [datetime]"'的成员,那么第一个单词将会被解析成列名,直到有空格为止:列名只是简单的"x"。
- ▶ isolation_level:获取或设置当前隔离级别。None 表示自动提交模式,或者可以是 "DEFERRED"、"IMMEDIATE"或"EXCLUSIVE"之一。
- ▶ in_transaction:如果为 True,则表示处于活动状态(有未提交的更改)。

在 sqlite3 模块中包含以下方法成员。

- ▶ sqlite3.connect(database [,timeout ,other optional arguments]):用于打开一个到 SQLite 数据库文件 database 的连接。可以使用 ":memory:" 在 RAM 中打开一个到 database 的数据库连接,而不是在磁盘上打开。如果数据库成功打开,则返回一个连接对象。当一个数据库被多个连接访问且其中一个修改了数据库时,此时 SQLite 数据库将被锁定,直到事务提交。参数 timeout 表示连接等待锁定的持续时间,直到发生异常断开连接。参数 timeout 的默认是 5.0(5 秒)。如果给定的数据库名 filename 不存在，则该调用将创建一个数据库。如果不想在当前目录中创建数据库，那么可以指定带有路径的文件名，这样就能在任意地方创建数据库。

- connection.cursor([cursorClass])：用于创建一个 cursor，将在 Python 数据库编程中用到。该方法接受一个单一的可选的参数 cursorClass。如果提供了该参数，则它必须是一个扩展自 sqlite3.Cursor 的自定义的 cursor 类。
- cursor.execute(sql [, optional parameters])：用于执行一个 SQL 语句。该 SQL 语句可以被参数化(即使用占位符代替 SQL 文本)。sqlite3 模块支持两种类型的占位符，即问号和命名占位符(命名样式)。例如：

```
cursor.execute("insert into people values (?, ?)", (who, age))
```

例如，在下面的实例文件 e.py 中，演示了使用方法 cursor.execute( )执行指定 SQL 语句的过程。

源码路径：codes\2\2-1\e.py

```python
import sqlite3

con = sqlite3.connect(":memory:")
cur = con.cursor()
cur.execute("create table people (name_last, age)")

who = "Yeltsin"
age = 72

# This is the qmark style:
cur.execute("insert into people values (?, ?)", (who, age))

# And this is the named style:
cur.execute("select * from people where name_last=:who and age=:age", {"who":
who, "age": age})

print(cur.fetchone())
```

执行后会输出：

```
('Yeltsin', 72)
```

- connection.execute(sql [, optional parameters])：是上面执行的由 cursor 对象提供的方法的快捷方式，通过调用 cursor 方法创建一个中间的 cursor 对象，然后通过给定的参数调用 cursor 的 execute 方法。
- cursor.executemany(sql, seq_of_parameters)：用于对 seq_of_parameters 中的所有参数或映射执行一个 SQL 命令。例如，在下面的实例文件 f.py 中，演示了使用方法 cursor.executemany( )执行指定 SQL 命令的过程。

源码路径：codes\2\2-1\f.py

```python
import sqlite3

class IterChars:
```

```python
    def __init__(self):
        self.count = ord('a')

    def __iter__(self):
        return self

    def __next__(self):
        if self.count > ord('z'):
            raise StopIteration
        self.count += 1
        return (chr(self.count - 1),) # this is a 1-tuple

con = sqlite3.connect(":memory:")
cur = con.cursor()
cur.execute("create table characters(c)")

theIter = IterChars()
cur.executemany("insert into characters(c) values (?)", theIter)

cur.execute("select c from characters")
print(cur.fetchall())
```

执行后会输出:

```
[('a',), ('b',), ('c',), ('d',), ('e',), ('f',), ('g',), ('h',), ('i',), ('j',),
('k',), ('l',), ('m',), ('n',), ('o',), ('p',), ('q',), ('r',), ('s',), ('t',),
('u',), ('v',), ('w',), ('x',), ('y',), ('z',)]
```

- ▶ connection.executemany(sql[, parameters]): 是一个由调用 cursor 方法创建的中间的光标对象的快捷方式, 然后通过给定的参数调用 cursor 的 executemany 方法。
- ▶ cursor.executescript(sql_script): 一旦接收到脚本就会执行多个 SQL 语句。首先执行 COMMIT 语句, 然后执行作为参数传入的 SQL 脚本。所有的 SQL 语句应该用分号";"分隔。例如, 在下面的实例文件 g.py 中, 演示了使用方法 cursor.executescript ( )执行多个 SQL 语句的过程。

源码路径: codes\2\2-1\g.py

```python
import sqlite3

con = sqlite3.connect(":memory:")
cur = con.cursor()
cur.executescript("""
    create table person(
        firstname,
        lastname,
        age
    );

    create table book(
        title,
        author,
        published
    );
```

```
    insert into book(title, author, published)
    values (
        'Dirk Gently''s Holistic Detective Agency',
        'Douglas Adams',
        1987
    );
""")
```

- connection.executescript(sql_script): 是一个由调用 cursor 方法创建的中间的光标对象的快捷方式，然后通过给定的参数调用光标的 executescript 方法。
- connection.total_changes( )：返回自数据库连接打开以来被修改、插入或删除的数据库总行数。
- connection.commit( )：用于提交当前的事务。如果未调用该方法，那么自上一次调用 commit( ) 以来所做的任何动作对其他数据库连接来说是不可见的。
- connection.create_function(name, num_params, func)：用于创建一个自定义的函数，随后可以在 SQL 语句中以函数名 name 来调用它。参数 num_params 表示此方法接受的参数数量(如果 num_params 为-1，函数可以取任意数量的参数)，参数 func 是一个可以被调用的 SQL 函数。例如，在下面的实例文件 b.py 中，演示了使用方法 create_function( )执行指定函数的过程。

源码路径：codes\2\2-1\b.py

```
import sqlite3
import hashlib

def md5sum(t):
    return hashlib.md5(t).hexdigest()

con = sqlite3.connect(":memory:")
con.create_function("md5", 1, md5sum)
cur = con.cursor()
cur.execute("select md5(?)", (b"foo",))
print(cur.fetchone()[0])
```

执行后会输出：

acbd18db4cc2f85cedef654fccc4a4d8

- connection.create_aggregate(name, num_params, aggregate_class)：用于创建一个用户定义的聚合函数。聚合类必须实现 step 方法，参数 num_params(如果 num_params 为-1，函数可以取任意数量的参数)表示该方法可以接受参数的数量。参数也可以是 finalize 方法，表示可以返回 SQLite 支持的任何类型，如 bytes、str、int、float 和 None。例如，在下面的实例文件 c.py 中，演示了使用方法 create_aggregate( ) 创建用户定义的聚合函数的过程。

源码路径：codes\2\2-1\c.py

```
import sqlite3
```

```python
class MySum:
    def __init__(self):
        self.count = 0

    def step(self, value):
        self.count += value

    def finalize(self):
        return self.count

con = sqlite3.connect(":memory:")
con.create_aggregate("mysum", 1, MySum)
cur = con.cursor()
cur.execute("create table test(i)")
cur.execute("insert into test(i) values (1)")
cur.execute("insert into test(i) values (2)")
cur.execute("select mysum(i) from test")
print(cur.fetchone()[0])
```

执行后会输出：

```
3
```

▶ connection.create_collation(name, callable)：功能是用指定的 name 和 callable 创建一个排序规则，会传递两个字符串参数给可调用对象。如果第一个比第二个小则返回-1，如果相等则返回 0；如果第一个比第二个大则返回 1。需要注意，可调用对象将会以 Python bytestring 的方式得到它的参数，一般为 UTF-8 编码。例如，在下面的实例文件 d.py 中，演示了使用方法 create_collation( )用自定义排序规则以"错误方式"进行排序的过程。

源码路径：codes\2\2-1\d.py

```python
import sqlite3

def collate_reverse(string1, string2):
    if string1 == string2:
        return 0
    elif string1 < string2:
        return 1
    else:
        return -1

con = sqlite3.connect(":memory:")
con.create_collation("reverse", collate_reverse)

cur = con.cursor()
cur.execute("create table test(x)")
cur.executemany("insert into test(x) values (?)", [("a",), ("b",)])
cur.execute("select x from test order by x collate reverse")
for row in cur:
    print(row)
con.close()
```

执行后会输出：

```
('b',)
('a',)
```

- connection.interrupt( )：从另一个线程中调用该方法来中止该连接正在执行的查询，查询会中止，调用者会得到一个异常。

- connection.set_authorizer(authorizer_callback)：用于注册一个回调。每当尝试访问数据库中表的列时会调用该回调，如果访问数据库被执行，回调应该返回 SQLITE_OK；如果 SQL 语句以错误中止，则回调应该返回 SQLITE_DENY；如果列被当成 NULL 值，回调应该返回 SQLITE_IGNORE。回调的第一个参数表示何种操作被授权。根据第一个参数，第二和第三个参数将提供回应或是 None。第四个参数是数据库的名称(main、temp 等)。第 5 个参数是最内部的触发器或视图的名字，它们负责访问请求；如果访问请求直接来自于输入的 SQL 代码则为 None。

- connection.set_progress_handler(handler, n)：用于注册一个回调。如果希望在长时间操作过程中从 SQLite 得到调用，这是非常有用的。如果希望清除之前安装的过程处理器，则以 None 为 handler 参数调用该方法。

- connection.set_trace_callback(trace_callback)：为 SQLite 后端实际执行的每个 SQL 语句调用寄存器 trace_callback。传递给回调的唯一参数是正在执行的语句(作为字符串)。

- connection.enable_load_extension(enabled)：允许或不允许 SQLite 引擎从共享库加载 SQLite 扩展。SQLite 扩展可以定义新的函数、聚合或全新的虚拟表实现。在默认情况下会禁用加载扩展。

- connection.load_extension(path)：用于从一个共享库加载 SQLite 扩展。在使用该方法之前必须用 enable_load_extension( )来允许扩展加载，在默认情况下禁用加载扩展。

- connection.connection.rollback( )：用于回滚自上一次调用 commit( ) 以来对数据库所做的更改。

- connection.close( )：用于关闭数据库连接。在此需要注意，它不会自动调用 commit( )。如果在之前未调用 commit( )方法，就会直接关闭数据库连接，之前所做的所有更改将全部丢失。

- cursor.fetchone( )：用于获取查询结果集中的下一行，返回一个单一的序列，当没有更多可用的数据时则返回 None。

- cursor.fetchmany([size=cursor.arraysize])：用于获取查询结果集中的下一行组，返回一个列表。当没有更多可用的行时，则返回一个空列表。该方法尝试获取由参数 size 指定的尽可能多的行。

- cursor.fetchall( )：用于获取查询结果集中所有(剩余)的行，返回一个列表。当没有可用的行时，则返回一个空列表。

- cursor.close( )：现在关闭光标(而不是每次调用 __del__ 时)，光标将从这一点向前不

- 可用。如果使用光标进行任何操作，则会出现 ProgrammingError 异常。
- register_converter(typename, callable)：功能是注册可调用对象，用来将来自数据库的字符串转换成为自定义的 Python 类型。在数据库中所有可支持的 typename 类型的值都可以调用该可调用对象。开发者在使用时需要注意：typename 的大小写和查询中类型的名称必须匹配。
- complete_statement(sql)：如果字符串 sql 包含一个或多个以分号结束的完整的 SQL 语句则返回 True。不会验证 SQL 的语法正确性，只是检查是否有未关闭的字符串常量以及语句是以分号结束。例如，在下面的实例文件 a.py 中，演示了使用方法 complete_statement(sql)生成一个 sqlite shell 的过程。

源码路径：codes\2\2-1\a.py

```python
import sqlite3

con = sqlite3.connect(":memory:")
con.isolation_level = None
cur = con.cursor()

buffer = ""

print("Enter your SQL commands to execute in sqlite3.")
print("Enter a blank line to exit.")

while True:
    line = input()
    if line == "":
        break
    buffer += line
    if sqlite3.complete_statement(buffer):
        try:
            buffer = buffer.strip()
            cur.execute(buffer)

            if buffer.lstrip().upper().startswith("SELECT"):
                print(cur.fetchall())
        except sqlite3.Error as e:
            print("An error occurred:", e.args[0])
        buffer = ""

con.close()
```

执行后会输出：

```
Enter your SQL commands to execute in sqlite3.
Enter a blank line to exit.
```

- enable_callback_tracebacks(flag)：在默认情况下不会在用户自定义的函数、聚合、转换器、授权者回调等地方得到回溯对象(调用栈对象)。如果想要调试它们，需要将参数 flag 设置为 True 然后调用此方法，之后可以在 sys.stderr 通过回调得到回溯。将参数 flag 设置为 False 后可以再次禁用该功能。

- Row.keys( )：用于返回列名称的列表。在查询后，它是在 Cursor.description 中每个元组的第一个成员。例如，在下面的实例文件 h.py 中，演示了使用 Row 对象和 keys( )方法的过程。

源码路径：codes\2\2-1\h.py

```
import sqlite3
conn = sqlite3.connect(":memory:")
c = conn.cursor()
c.execute('''create table stocks
(date text, trans text, symbol text,
 qty real, price real)''')
c.execute("""insert into stocks
        values ('2018-01-05','BUY','RHAT',100,35.14)""")
conn.commit()
c.close()

conn.row_factory = sqlite3.Row
c = conn.cursor()
print(c.execute('select * from stocks'))
r = c.fetchone()
print(type(r))
print(tuple(r))
print(len(r))
print(r[2])
print(r.keys())
print(r['qty'])
for member in r:
    print(member)
```

执行后会输出：

```
<sqlite3.Cursor object at 0x0000021B3B291F10>
<class 'sqlite3.Row'>
('2018-01-05', 'BUY', 'RHAT', 100.0, 35.14)
5
RHAT
['date', 'trans', 'symbol', 'qty', 'price']
100.0
2018-01-05
BUY
RHAT
100.0
35.14
```

2.1.2 使用 sqlite3 模块操作 SQLite 3 数据库

根据 DB-API 2.0 规范规定，Python 语言操作 SQLite 3 数据库的基本流程如下。

(1) 导入相关库或模块(sqlite3)。

(2) 使用 connect( )连接数据库，并获取数据库连接对象。

(3) 使用 con.cursor( )获取游标对象。

(4) 使用游标对象的方法(execute( )、executemany( )、fetchall( )等)操作数据库,实现插入、修改和删除操作,并查询获取显示相关的记录。在 Python 程序中,连接函数 sqlite3.connect( )有以下两个常用参数。

- database:表示要访问的数据库名。
- timeout:表示访问数据的超时设定。

其中,参数 database 表示用字符串的形式指定数据库的名称,如果数据库文件位置不是当前目录,则必须要写出其相对路径或绝对路径。还可以用:memory:表示使用临时放入内存的数据库。当退出程序时,数据库中的数据也就不存在了。

(5) 用 close( )方法关闭游标对象和数据库连接。数据库操作完成后,必须及时调用其 close( )方法关闭数据库连接,这样做的目的是减轻数据库服务器的压力。

例如,在下面的实例文件 sqlite.py 中,演示了使用 sqlite3 模块操作 SQLite 3 数据库的过程。

源码路径:codes\2\2-1\sqlite.py

```
import sqlite3                    #导入内置模块
import random                     #导入内置模块
#初始化变量src,设置用于随机生成字符串中的所有字符
src = 'abcdefghijklmnopqrstuvwxyz'
def get_str(x,y):                 #生成字符串函数get_str()
    str_sum = random.randint(x,y) #生成x和y之间的随机整数
    astr = ''                     #为变量astr赋值
    for i in range(str_sum):      #遍历随机数
        astr += random.choice(src)#累计求和生成的随机数
    return astr                   #返回和
def output():                     #函数output()用于输出数据库表中的所有信息
    cur.execute('select * from biao')#查询表biao中的所有信息
    for sid,name,ps in cur:       #查询表中的3个字段sid、name和ps
        print(sid,' ',name,' ',ps)  #显示3个字段的查询结果

def output_all():                 #函数output_all()用于输出数据库表中的所有信息
    cur.execute('select * from biao')#查询表biao中的所有信息
    for item in cur.fetchall():   #获取查询到的所有数据
        print(item)               #打印显示获取到的数据

def get_data_list(n):             #函数get_data_list()用于生成查询列表
    res = []                      #列表初始化
    for i in range(n):            #遍历列表
        res.append((get_str(2,4),get_str(8,12)))#生成列表
    return res                    #返回生成的列表
if __name__ == '__main__':
    print("建立连接...")           #打印提示
    con = sqlite3.connect(':memory:')#开始建立和数据库的连接
    print("建立游标...")
    cur = con.cursor()            #获取游标
    print('创建一张表biao...')     #打印提示信息
    #在数据库中创建表biao,设置了表中的各个字段
```

```
    cur.execute('create table biao(id integer primary key autoincrement not
null,name text,passwd text)')
    print('插入一条记录...')              #打印提示信息
    #插入一条数据信息
    cur.execute('insert into biao
(name,passwd)values(?,?)',(get_str(2,4),get_str(8,12),))
    print('显示所有记录...')              #打印提示信息
    output()                              #显示数据库中的数据信息
    print('批量插入多条记录...')          #打印提示信息
    #插入多条数据信息
    cur.executemany('insert into biao
(name,passwd)values(?,?)',get_data_list(3))
    print("显示所有记录...")              #打印提示信息
    output_all()                          #显示数据库中的数据信息
    print('更新一条记录...')              #打印提示信息
    #修改表biao中的一条信息
    cur.execute('update biao set name=? where id=?',('aaa',1))
    print('显示所有记录...')              #打印提示信息
    output()                              #显示数据库中的数据信息
print('删除一条记录...')                  #打印提示信息
    #删除表biao中的一条数据信息
    cur.execute('delete from biao where id=?',(3,))
    print('显示所有记录：')               #打印提示信息
    output()                              #显示数据库中的数据信息
```

在上述实例代码中，首先定义了两个能够生成随机字符串的函数，生成的随机字符串作为数据库中存储的数据。然后定义 output( )和 output-all( )方法，功能是分别通过遍历 cursor、调用 cursor 的方式来获取数据库表中的所有记录并输出。然后在主程序中，依次通过建立连接，获取连接的 cursor，通过 cursor 的 execute( )和 executemany( )等方法来执行 SQL 语句，以实现插入一条记录、插入多条记录、更新记录和删除记录的功能。最后依次关闭游标和数据库连接。执行后会输出：

```
建立连接...
建立游标...
创建一张表biao...
插入一条记录...
显示所有记录...
1    bld    zbynubfxt
批量插入多条记录...
显示所有记录...
(1, 'bld', 'zbynubfxt')
(2, 'owd', 'lqpperrey')
(3, 'vc', 'fqrbarwsotra')
(4, 'yqk', 'oyzarvrv')
更新一条记录...
显示所有记录...
1    aaa    zbynubfxt
2    owd    lqpperrey
3    vc     fqrbarwsotra
4    yqk    oyzarvrv
删除一条记录...
```

```
显示所有记录:
1   aaa   zbynubfxt
2   owd   lqpperrey
4   yqk   oyzarvrv
```

2.1.3　SQLite 和 Python 的类型

SQLite 可以支持的类型有 NULL、INTEGER、REAL、TEXT 和 BLOB，所以表 2-1 中的 Python 类型可以直接发送给 SQLite。

表 2-1　SQLite 可以直接使用的 Python 的类型

Python 类型	SQLite 类型
None	NULL
int	INTEGER
float	REAL
str	TEXT
bytes	BLOB

在默认情况下，SQLite 将表 2-2 中的类型转换成 Python 类型。

表 2-2　转换成 Python 类型

SQLite 类型	Python 类型
NULL	None
INTEGER	int
REAL	float
TEXT	在默认情况下取决于 text_factory 和 str
BLOB	bytes

在 SQLite 处理 Python 数据的过程中，有可能需要处理其他更多种类型的数据，而这些数据类型 SQLite 并不支持，此时需要用到类型扩展技术来实现所需功能。在 Python 语言的 sqlite3 模块中，其类型系统可以用两种方式来扩展数据类型：通过对象适配，可以在 SQLite 数据库中存储其他的 Python 类型；通过转换器让 sqlite3 模块将 SQLite 类型转成不同的 Python 类型。

1. 使用适配器来存储额外的 Python 类型的 SQLite 数据库

因为 SQLite 只支持有限的类，要使用其他 Python 类型与 SQLite 进行交互，就必须适应它们为 sqlite3 模块支持的 SQLite 类型之一，包括 NoneType、int、float、str 或 bytes。通过使用以下两种方法，可以使 sqlite3 模块适配一个 Python 类型到一个支持的类型。

1）编写类进行适应

开发者可以编写一个自定义类，假设编写了以下一个类：

```
class Point:
    def __init__(self, x, y):
        self.x, self.y = x, y
```

想要在某个 SQLite 列中存储类 Point,首先得选择一个支持的类型,这个类型可以用来表示 Point。假定使用 str,并用分号来分隔坐标。需要给类加一个 __conform__(self, protocol) 方法,该方法必须返回转换后的值。参数 protocol 为 PrepareProtocol 类型。例如,在下面的实例文件 i.py 中,演示了将自定义类 Point 适配 SQLite3 数据库的过程。

源码路径:codes\2\2-1\i.py

```
import sqlite3

class Point:
    def __init__(self, x, y):
        self.x, self.y = x, y

def adapt_point(point):
    return "%f;%f" % (point.x, point.y)

sqlite3.register_adapter(Point, adapt_point)

con = sqlite3.connect(":memory:")
cur = con.cursor()

p = Point(4.0, -3.2)
cur.execute("select ?", (p,))
print(cur.fetchone()[0])
```

执行后会输出:

4.000000;-3.200000

2) 注册可调用的适配器

例如,有一种可能性是创建一个函数,用来将类型转换成字符串表现形式,然后使用函数 register_adapter( ) 来注册该函数。例如,在下面的实例文件 j.py 中,演示了使用函数 register_adapter( ) 注册适配器函数的过程。

源码路径:codes\2\2-1\j.py

```
import sqlite3

class Point:
    def __init__(self, x, y):
        self.x, self.y = x, y

def adapt_point(point):
    return "%f;%f" % (point.x, point.y)

sqlite3.register_adapter(Point, adapt_point)

con = sqlite3.connect(":memory:")
cur = con.cursor()
```

```
p = Point(4.0, -3.2)
cur.execute("select ?", (p,))
print(cur.fetchone()[0])
```

执行后会输出:

```
4.000000;-3.200000
```

在sqlite3模块中,为Python内置的datetime.date和datetime.datetime类型设置了两个默认适配器。假设想要将datetime.datetime对象不以ISO形式存储,而是保存成Unix时间戳,则可以通过下面的实例文件k.py实现。

源码路径:codes\2\2-1\k.py

```
import sqlite3
import datetime
import time

def adapt_datetime(ts):
    return time.mktime(ts.timetuple())

sqlite3.register_adapter(datetime.datetime, adapt_datetime)

con = sqlite3.connect(":memory:")
cur = con.cursor()

now = datetime.datetime.now()
cur.execute("select ?", (now,))
print(cur.fetchone()[0])
```

执行后会输出:

```
1513239416.0
```

2. 将自定义 Python 类型转换成 SQLite 类型

在 Python 程序中,可以编写适配器将自定义 Python 类型转换成 SQLite 类型。再次以前面的 Point 类进行举例,假设在 SQLite 中以字符串的形式存储以分号分隔的 x、y 坐标。可以先定义以下转换器函数 convert_point( ),用于接收字符串参数,并从中构造一个 Point 对象。转换器函数总是使用 bytes 对象调用,而无论将数据类型发送到 SQLite 的哪种数据类型。

```
def convert_point(s):
    x, y = map(float, s.split(b";"))
    return Point(x, y)
```

接下来需要让 sqlite3 模块知道从数据库中实际选择的是一个点,可以通过以下两种方法实现这个功能。

▶ 隐式地通过声明的类型。
▶ 显式地通过列名。

例如,下面的实例文件 l.py 演示了上述两种方法的实现过程。

源码路径：codes\2\2-1\l.py

```python
import sqlite3
class Point:
    def __init__(self, x, y):
        self.x, self.y = x, y

    def __repr__(self):
        return "(%f;%f)" % (self.x, self.y)

def adapt_point(point):
    return ("%f;%f" % (point.x, point.y)).encode('ascii')

def convert_point(s):
    x, y = list(map(float, s.split(b";")))
    return Point(x, y)

# Register the adapter
sqlite3.register_adapter(Point, adapt_point)

# Register the converter
sqlite3.register_converter("point", convert_point)

p = Point(4.0, -3.2)

#########################
# 1) Using declared types
con = sqlite3.connect(":memory:", detect_types=sqlite3.PARSE_DECLTYPES)
cur = con.cursor()
cur.execute("create table test(p point)")

cur.execute("insert into test(p) values (?)", (p,))
cur.execute("select p from test")
print("with declared types:", cur.fetchone()[0])
cur.close()
con.close()

#######################
# 1) Using column names
con = sqlite3.connect(":memory:", detect_types=sqlite3.PARSE_COLNAMES)
cur = con.cursor()
cur.execute("create table test(p)")

cur.execute("insert into test(p) values (?)", (p,))
cur.execute('select p as "p [point]" from test')
print("with column names:", cur.fetchone()[0])
cur.close()
con.close()
```

执行后会输出：

```
with declared types: (4.000000;-3.200000)
with column names: (4.000000;-3.200000)
```

3. 默认适配器和转换器

在 Python 语言的 datetime 模块中，有对 date 和 datetime 类型默认的适配器，将 ISO dates/ISO timestamps 发送给 SQLite。默认的转换器以 date 为名注册给 datetime.date，以 timestamp 为名注册给 datetime.datetime。这样在大多数情况下可以在 Python 中使用 date/timestamp 时不需要额外的动作，适配器的格式兼容于 SQLite 的 date/time 函数。

例如，在下面的实例文件 m.py 中，演示了使用默认适配器和转换器的过程。

 源码路径：codes\2\2-1\m.py

```python
import sqlite3
import datetime

con = sqlite3.connect(":memory:",
detect_types=sqlite3.PARSE_DECLTYPES|sqlite3.PARSE_COLNAMES)
cur = con.cursor()
cur.execute("create table test(d date, ts timestamp)")

today = datetime.date.today()
now = datetime.datetime.now()

cur.execute("insert into test(d, ts) values (?, ?)", (today, now))
cur.execute("select d, ts from test")
row = cur.fetchone()
print(today, "=>", row[0], type(row[0]))
print(now, "=>", row[1], type(row[1]))

cur.execute('select current_date as "d [date]", current_timestamp as "ts [timestamp]"')
row = cur.fetchone()
print("current_date", row[0], type(row[0]))
print("current_timestamp", row[1], type(row[1]))
```

执行后会输出：

```
2017-12-14 => 2017-12-14 <class 'datetime.date'>
2017-12-14 16:37:55.125396 => 2017-12-14 16:37:55.125396 <class 'datetime.datetime'>
current_date 2017-12-14 <class 'datetime.date'>
current_timestamp 2017-12-14 08:37:55 <class 'datetime.datetime'>
```

> **注意**
> 如果存储在 SQLite 中的时间戳的小数部分大于 6 个数字，它的值将由时间戳转换器截断至微秒的精度。

2.2 操作 MySQL 数据库

在 Python 3.x 版本中,使用内置库 PyMySQL 来连接 MySQL 数据库服务器，Python 2 版本中使用库 mysqldb。PyMySQL 完全

扫码观看本节视频讲解

遵循 Python 数据库 API v2.0 规范，并包含了 pure-Python MySQL 客户端库。在本节的内容中，将详细讲解在 Python 程序中操作 MySQL 数据库的知识。

2.2.1 搭建 PyMySQL 环境

在使用 PyMySQL 之前，必须先确保已经安装 PyMySQL。PyMySQL 的下载地址是 https://github.com/PyMySQL/PyMySQL。如果还没有安装，可以使用以下命令安装最新版的 PyMySQL：

```
pip install PyMySQL
```

安装成功后的界面效果如图 2-1 所示。

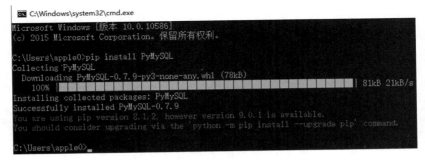

图 2-1　CMD 界面

如果当前系统不支持 pip 命令，可以使用以下两种方式进行安装。

(1) 使用 git 命令下载安装包安装：

```
$ git clone https://github.com/PyMySQL/PyMySQL
$ cd PyMySQL/
$ python3 setup.py install
```

(2) 如果需要指定版本号，可以使用 curl 命令进行安装：

```
$ # X.X 为 PyMySQL 的版本号
$ curl -L https://github.com/PyMySQL/PyMySQL/tarball/pymysql-X.X | tar xz
$ cd PyMySQL*
$ python3 setup.py install
$ # 现在可以删除 PyMySQL* 目录
```

> **注　意**
> 必须确保拥有 root 权限才可以安装上述模块。另外，在安装过程中可能会出现 "ImportError: No module named setuptools" 的错误提示，这个提示的意思是没有安装 setuptools，可以访问 https://pypi.python.org/pypi/setuptools 找到各个系统的安装方法。如在 Linux 系统中的安装实例如下：
> ```
> $ wget https://bootstrap.pypa.io/ez_setup.py
> $ python3 ez_setup.py
> ```

2.2.2 实现数据库连接

在连接数据库之前，需按照以下步骤进行操作。
(1) 安装 MySQL 数据库和 PyMySQL。
(2) 在 MySQL 数据库中创建数据库 TESTDB。
(3) 在 TESTDB 数据库中创建表 EMPLOYEE。
(4) 在表 EMPLOYEE 中分别添加 5 个字段，分别是 FIRST_NAME、LAST_NAME、AGE、SEX 和 INCOME。在 MySQL 数据库，表 EMPLOYEE 的界面效果如图 2-2 所示。

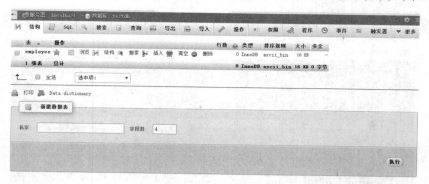

图 2-2　表 EMPLOYEE 的界面效果

(5) 假设本地 MySQL 数据库的登录用户名为 root，密码为 66688888。例如，在下面的实例文件 mysql.py 中，演示了显示 PyMySQL 数据库版本号的过程。

源码路径：daima\2\2-2\mysql.py

```python
import pymysql
#打开数据库连接
db = pymysql.connect("localhost","root","66688888","TESTDB" )
#使用 cursor()方法创建一个游标对象 cursor
cursor = db.cursor()
#使用 execute()方法执行 SQL 查询
cursor.execute("SELECT VERSION()")
#使用 fetchone() 方法获取单条数据
data = cursor.fetchone()
print ("Database version : %s " % data)
#关闭数据库连接
db.close()
```

执行后会输出：

```
Database version : 5.7.17-log
```

2.2.3 创建数据库表

在 Python 程序中，可以使用方法 execute( )在数据库中创建一个新表。例如，在下面的实例文件 new.py 中，演示了在 PyMySQL 数据库中创建新表 EMPLOYEE 的过程。

源码路径：daima\3\2-2\new.py

```python
import pymysql
#打开数据库连接
db = pymysql.connect("localhost","root","66688888","TESTDB" )
#使用 cursor()方法创建一个游标对象 cursor
cursor = db.cursor()
#使用 execute() 方法执行 SQL，如果表存在则删除
cursor.execute("DROP TABLE IF EXISTS EMPLOYEE")
#使用预处理语句创建表
sql = """CREATE TABLE EMPLOYEE (
         FIRST_NAME  CHAR(20) NOT NULL,
         LAST_NAME  CHAR(20),
         AGE INT,
         SEX CHAR(1),
         INCOME FLOAT )"""
cursor.execute(sql)
#关闭数据库连接
db.close()
```

执行上述代码后，将在 MySQL 数据库中创建一个名为 EMPLOYEE 的新表，执行后的效果如图 2-3 所示。

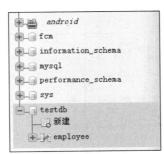

图 2-3 执行后的效果

2.3 使用 MariaDB 数据库

MariaDB 是一种开源数据库，是 MySQL 数据库的一个分支。因为某些历史原因，有不少用户担心 MySQL 数据库会停止开源，所以 MariaDB 逐步发展成为 MySQL 替代品的数据库工具之一。本节将详细讲解使用 MySQL 第三方库来操作 MariaDB 数据库的知识。

扫码观看本节视频讲解

2.3.1 搭建 MariaDB 数据库环境

作为一款经典的关系数据库产品，搭建 MariaDB 数据库环境的基本流程如下。

(1) 登录 MariaDB 官网下载页面 https://downloads.mariadb.org/，如图 2-4 所示。

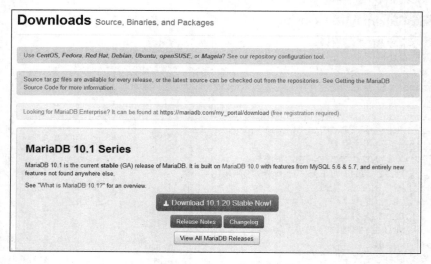

图 2-4　MariaDB 官网下载页面

(2) 单击 Download 10…按钮弹出下载界面，如图 2-5 所示。在此需要根据计算机系统的版本进行下载，如笔者的计算机是 64 位的 Windows 10 系统，所以选择 mariadb-10.1.20-winx64.msi 进行下载。

File Name	Package Type	OS / CPU	Size	Meta
mariadb-10.1.20.tar.gz	source tar.gz file	Source	61.3 MB	MD5 SHA1 Signature Instructions
mariadb-10.1.20-winx64.msi	MSI Package	Windows x86_64	160.7 MB	MD5 SHA1 Signature Instructions
mariadb-10.1.20-winx64.zip	ZIP file	Windows x86_64	333.7 MB	MD5 SHA1 Signature Instructions
mariadb-10.1.20-win32.msi	MSI Package	Windows x86	156.9 MB	MD5 SHA1 Signature Instructions
mariadb-10.1.20-win32.zip	ZIP file	Windows x86	330.2 MB	MD5 SHA1 Signature Instructions
mariadb-10.1.20-linux-glibc_214-x86_64.tar.gz (requires GLIBC_2.14+)	gzipped tar file	Linux x86_64	476.6 MB	MD5 SHA1 Signature Instructions

图 2-5　下载页面

(3) 下载完成后会得到一个安装文件 mariadb-10.1.20-winx64.msi，双击这个文件后弹出欢迎安装对话框，如图 2-6 所示。

(4) 单击 Next 按钮后弹出用户协议对话框，在此勾选 I accept…复选框，如图 2-7 所示。

(5) 单击 Next 按钮后弹出典型设置对话框，在此设置程序文件的安装路径，如图 2-8 所示。

(6) 单击 Next 按钮后弹出设置密码对话框，在此设置管理员用户 root 的密码，如图 2-9

所示。

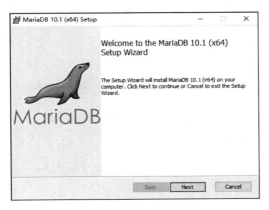

图 2-6　欢迎安装对话框　　　　　图 2-7　用户协议对话框

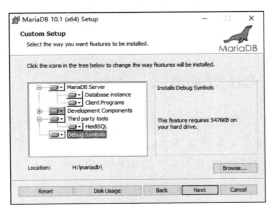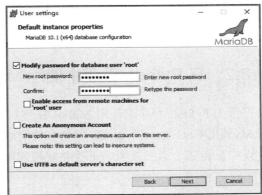

图 2-8　典型设置对话框　　　　　图 2-9　设置密码对话框

(7) 单击 Next 按钮后弹出默认实例属性对话框，在此设置服务器名字和 TCP 端口号，如图 2-10 所示。

(8) 单击 Next 按钮弹出准备安装对话框，如图 2-11 所示。

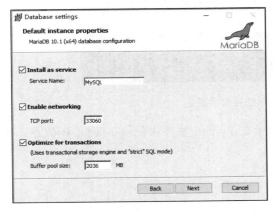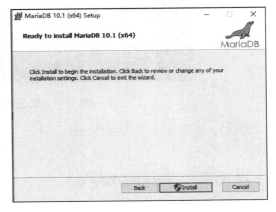

图 2-10　默认实例属性对话框　　　　　图 2-11　准备安装对话框

(9) 单击 Install 按钮后弹出安装进度条界面，开始安装 MariaDB，如图 2-12 所示。

(10) 安装进度完成后弹出完成安装对话框，单击 Finish 按钮后完成安装，如图 2-13 所示。

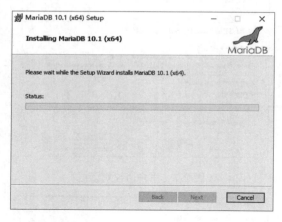
图 2-12　安装进度条界面

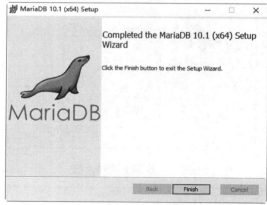
图 2-13　完成安装对话框

2.3.2　在 Python 程序中使用 MariaDB 数据库

当在 Python 程序中使用 MariaDB 数据库时，需要在程序中加载 Python 语言的第三方库 MySQL Connector Python。但是在使用这个第三方库操作 MariaDB 数据库之前，需要先下载并安装这个第三方库。下载并安装的过程非常简单，只需在控制台中执行以下命令即可实现：

```
pip install mysql-connector
```

安装成功时的界面效果如图 2-14 所示。

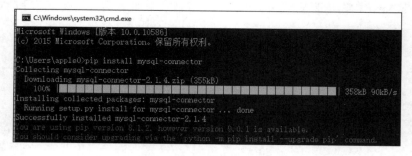
图 2-14　第三方库下载并安装成功

例如，在下面的实例文件 md.py 中，演示了在 Python 程序中使用 MariaDB 数据库的过程。

源码路径：codes\2\2-3\md.py

```
from mysql import connector
import random                              #导入内置模块
…省略部分代码…
```

```python
if __name__ == '__main__':
    print("建立连接...")                    #打印显示提示信息
    #建立数据库连接
    con = connector.connect(user='root',password=
                    '66688888',database='md')
    print("建立游标...")                    #打印显示提示信息
    cur = con.cursor()                      #建立游标
    print('创建一张表mdd...')               #打印显示提示信息
    #创建数据库表mdd
    cur.execute('create table mdd(id int primary key auto_increment not null,name text,passwd text)')
    #在表mdd中插入一条数据
    print('插入一条记录...')                #打印显示提示信息
    cur.execute('insert into mdd(name,passwd) values(%s,%s)',(get_str(2,4),get_str(8,12),))
    print('显示所有记录...')                #打印显示提示信息
    output()                                #显示数据库中的数据信息
    print('批量插入多条记录...')            #打印显示提示信息
    #在表mdd中插入多条数据
    cur.executemany('insert into mdd(name,passwd) values(%s,%s)',get_data_list(3))
    print("显示所有记录...")                #打印显示提示信息
    output_all()                            #显示数据库中的数据信息
    print('更新一条记录...')                #打印显示提示信息
    #修改表mdd中的一条数据
    cur.execute('update mdd set name=%s where id=%s',('aaa',1))
    print('显示所有记录...')                #打印显示提示信息
    output()                                #显示数据库中的数据信息
    print('删除一条记录...')                #打印显示提示信息
    #删除表mdd中的一条数据信息
    cur.execute('delete from mdd where id=%s',(3,))
    print('显示所有记录：')                 #打印显示提示信息
    output()                                #显示数据库中的数据信息
```

在上述实例代码中，使用 mysql-connector-python 模块中的函数 connect( )建立了和 MariaDB 数据库的连接。连接函数 connect( )在 mysql.connector 中定义，此函数的语法原型如下：

```
connect(host, port,user, password, database, charset)
```

- host：访问数据库的服务器主机(默认为本机)。
- port：访问数据库的服务端口(默认为3306)。
- user：访问数据库的用户名。
- password：访问数据库用户名的密码。
- database：访问数据库名称。
- charset：字符编码(默认为uft8)。

执行后将显示创建数据表并实现数据插入、更新和删除操作的过程。执行后会输出：

```
建立连接...
建立游标...
创建一张表mdd...
```

```
插入一条记录...
显示所有记录...
1   kpv    lrdupdsuh
批量插入多条记录...
显示所有记录...
(1, 'kpv', 'lrdupdsuh')
(2, 'hsue', 'ilrleakcoh')
(3, 'hb', 'dzmcajvm')
(4, 'll', 'ngjhixta')
更新一条记录...
显示所有记录...
1   aaa    lrdupdsuh
2   hsue   ilrleakcoh
3   hb     dzmcajvm
4   ll     ngjhixta
删除一条记录...
显示所有记录:
1   aaa    lrdupdsuh
2   hsue   ilrleakcoh
4   ll     ngjhixta
```

> **注意**
>
> 在操作 MariaDB 数据库时，与操作 SQLite3 的 SQL 语句不同的是，SQL 语句中的占位符不是 ?，而是 %s。

2.3.3　使用 MariaDB 创建 MySQL 数据库

请看下面的实例文件 123.py，功能是使用 MariaDB 创建一个指定的 MySQL 数据库。

源码路径：codes\2\2-3\123.py

```python
import socket
import time
import pymysql as mariadb

# 设置要链接的数据库的名字

DB_NAME='mariadb'

# 如果您已经注册了，请先登录
TABLES = {}

TABLES['location'] = (
   "CREATE TABLE IF NOT EXISTS 'location' ("
   "  'id' int(255) NOT NULL AUTO_INCREMENT,"
   "  'latitud' varchar(15) NOT NULL,"
   "  'longitud' varchar(15) NOT NULL,"
   "  'Fecha' varchar(22) NOT NULL,"
   "  'Hora' varchar(22) NOT NULL,"
   "  PRIMARY KEY ('id'), UNIQUE KEY 'Hora' ('Hora')"
   ") ENGINE=InnoDB")

# 使用用户名和密码连接服务器
```

```python
cnx = mariadb.connect(host='localhost', user='root', password='66688888')
cursor = cnx.cursor()

# 创建数据库
def generate_database(curs):
    try:
        # 删除已存在的数据库
        # curs.execute("DROP DATABASE IF EXISTS {}".format(DB_NAME))
        # 如果数据库不存在则创建它
        curs.execute(
            "CREATE DATABASE IF NOT EXISTS {} DEFAULT CHARACTER SET 'utf8'".format(DB_NAME))
    except mariadb.Error as err:
        print("Failed creating database: {}".format(err))
        exit(1)
    else:
        print("Database OK")

try:
    generate_database(cursor)
except mariadb.Error as err:
    print("Error: {}".format(err))

cursor.execute("USE {}".format(DB_NAME))
for name, ddl in TABLES.items():
    try:
        print("Creating table {}: ".format(name), end='')
        cursor.execute(ddl)
    except mariadb.Error as err:
        print("Failed creating table: {}".format(err))
        exit(1)
    else:
        print("Table OK")

cnx.commit()
cnx.close()

def main():
    # 创建 TCP/IP socket
    sock = socket.socket(socket.AF_INET, socket.SOCK_DGRAM)
    HOST = socket.gethostbyname(socket.gethostname())
    PORT = 10
    # 绑定 socket 端口

    server_address = (HOST, PORT)
    print('Inicializando en Host IPV4 %s Puerto %s' % server_address)
    sock.bind(server_address)

    while True:
        try:
            while True:
                print("Connected")
                raw_data, addr = sock.recvfrom(65535)
                save_data = str(raw_data)[2:]
                # 区分纬度和经度的格式,分别打印经度和纬度信息,这样可以实现标准本地化功能
```

```python
                    if raw_data:
                        # print('recibido ' + save_data)
                        op, evento, fecha, lat, lon = obtMsg(save_data);
                        # print(op, evento)
                        if op:
                            print('Evento: ' + str(evento) + ', ' + 'la latitud es: ' + str(lat) + ' y la longitud es: ' + str(lon))
                            print('Fecha del dato: ' + fecha)
                            # 连接数据库
                            cnx = mariadb.connect(host='localhost', user='root', password='66688888')
                            cursor = cnx.cursor()
                            cursor.execute("USE {}".format(DB_NAME))
                            # 插入数据
                            add_location = ("INSERT INTO location "
                                            "(latitud, longitud, Fecha ,Hora) "
                                            "VALUES (%s, %s, %s)")
                            data_location = (str(lat), str(lon), fecha, hora)
                            # 插入新的本地化信息
                            cursor.execute(add_location, data_location)
                            cnx.commit()
                            cursor.close()
                            cnx.close()

                        else:
                            print("*********************************************")
                            print(" Mensaje Ignorado ")
                            print("*********************************************")
                    else:
                        break
        finally:
            print("No se estan recibiendo más datos")

def obtMsg(d):
    # 注意 REV 和 RPV 之间的区别
    if d[0:4] == ">REV":
        op = True
        # 用于打印数据(作为确认信息)
        evento = int(d[4:6])
        # Se almacenan los index de eventos
        fecha = obtFecha(d[6:10], d[10], d[11:16])
        # 除日期存储为 String 类型
        lat = float(d[17:19]) + (float(d[19:24]) / 100000)
        if d[16] == "-":
            lat = -lat
        lon = float(d[25:28]) + (float(d[28:33]) / 100000)
        if d[24] == "-":
            lon = -lon
    else:
        op = False
        evento = 0
        fecha = ' '
        lat = 0
        lon = 0
    return op, evento, fecha, lat, lon
```

```
def obtFecha(sem,dia,hora):
    seg = int(sem) * 7 * 24 * 60 * 60 + (int(dia) + 3657) * 24 * 60 * 60 + int(hora) * 5 * 60 * 60
    # 将数字(以秒计)转换为日期格式 %b %d %Y %M %S
    # 具体参考(Vease https://docs.python.org/2/library/time.html)
    # t = time.mktime(seg)
    fecha = time.strftime("%b %d %Y %H:%M:%S", time.localtime(seg))
    return fecha

main()
```

执行后会创建 MySQL 数据库 mariadb，在此数据库中创建一个名为 location 的表。执行后会输出：

```
Database OK
Creating table location: Table OK
Inicializando en Host IPV4 192.168.1.102 Puerto 10
Connected
```

2.4 使用 MongoDB 数据库

MongoDB 是一个基于分布式文件存储的数据库，由 C++语言编写，旨在为 Web 应用提供可扩展的高性能数据存储解决方案。MongoDB 是一个介于关系数据库和非关系数据库之间的产品，是非关系数据库中功能最丰富、最像关系数据库的。在本节的内容中，将详细讲解在 Python 程序中使用 MongoDB 数据库的知识。

扫码观看本节视频讲解

2.4.1 搭建 MongoDB 环境

（1）在 MongoDB 官网中提供了可用于 32 位和 64 位系统的预编译二进制包，读者可以从 MongoDB 官网下载安装包，下载地址是 https://www.mongodb.com/download-center#enterprise，如图 2-15 所示。

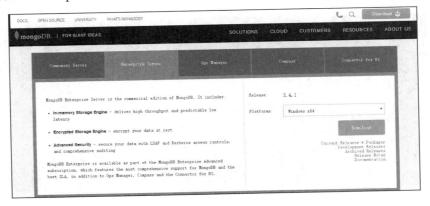

图 2-15 MongoDB 下载页面

(2) 根据当前计算机的操作系统选择下载安装包，因为笔者是 64 位的 Windows 系统，所以选择 Windows x64，然后单击 Download 按钮。在弹出的界面中选择 msi，如图 2-16 所示。

(3) 下载完成后得到一个 .msi 格式文件，双击该文件，然后按照操作提示进行安装即可。安装界面如图 2-17 所示。

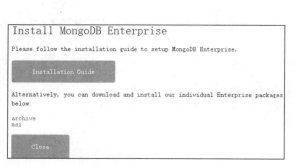
图 2-16　选择 msi 文件

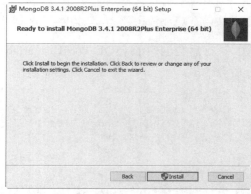
图 2-17　安装界面

2.4.2　在 Python 程序中使用 MongoDB 数据库

在 Python 程序中使用 MongoDB 数据库时，必须首先确保安装了 pymongo 这一第三方库。如果下载的是 exe 格式的安装文件，则可以直接运行安装。如果是压缩包的安装文件，可以使用以下命令进行安装：

```
pip install pymongo
```

如果没有下载安装文件，可以通过以下命令进行在线安装：

```
easy_install pymongo
```

安装完成后的界面效果如图 2-18 所示。

图 2-18　安装完成后的界面效果

例如，在下面的实例文件 mdb.py 中，演示了在 Python 程序中使用 MongoDB 数据库的过程。

源码路径：codes\2\2-4\mdb.py

```python
from pymongo import MongoClient
import random
...省略部分代码...
if __name__ == '__main__':
    print("建立连接...")                    #打印提示信息
    stus = MongoClient().test.stu           #建立连接
    print('插入一条记录...')                #打印提示信息
    #向表 stu 中插入一条数据
    stus.insert({'name':get_str(2,4),'passwd':get_str(8,12)})
    print("显示所有记录...")                #打印提示信息
    stu = stus.find_one()                   #获取数据库信息
    print(stu)                              #显示数据库中的数据信息
    print('批量插入多条记录...')            #打印提示信息
    stus.insert(get_data_list(3))           #向表 stu 中插入多条数据
    print('显示所有记录...')                #打印提示信息
    for stu in stus.find():                 #遍历数据信息
        print(stu)                          #显示数据库中的数据信息
    print('更新一条记录...')                #打印提示信息
    name = input('请输入记录的name:')#提示输入要修改的数据
    #修改表 stu 中的一条数据
    stus.update({'name':name},{'$set':{'name':'langchao'}})
    print('显示所有记录...')                #打印提示信息
    for stu in stus.find():                 #遍历数据
        print(stu)                          #显示数据库中的数据信息
    print('删除一条记录...')                #打印提示信息
    name = input('请输入记录的name:')#提示输入要删除的数据
    stus.remove({'name':name})              #删除表中的数据
    print('显示所有记录...')                #打印提示信息
    for stu in stus.find():                 #遍历数据信息
        print(stu)                          #显示数据库中的数据信息
```

在上述实例代码中使用两个函数生成字符串。在主程序中首先连接集合，然后使用集合对象的方法对集合中的文档进行插入、更新和删除操作。每当数据被修改后，会显示集合中所有文档，以验证操作结果的正确性。

在运行本实例时，初学者很容易遇到以下 Mongo 运行错误：

```
Failed to connect 127.0.0.1:27017,reason:errno:10061 由于目标计算机积极拒绝，无法连接...
```

发生上述错误的原因是没有开启 MongoDB 服务，下面是开启 MongoDB 服务的命令：

```
mongod --dbpath "h:\data"
```

在上述命令中，"h:\data" 是一个保存 MongoDB 数据库数据的目录，读者可以随意在本地计算机硬盘中创建，并且还可以自定义目录名字。在 CMD 控制台界面中，开启 MongoDB 服务成功时的界面效果如图 2-19 所示。

图 2-19 开启 MongoDB 服务成功时的界面效果

在运行本实例程序时，必须在 CMD 控制台中启动 MongoDB 服务，并且确保上述控制台界面处于打开状态。本实例执行后会输出：

```
建立连接...
插入一条记录...
显示所有记录...
{'_id': ObjectId('586243795cd071f570ed3b39'), 'name': 'vvtj', 'passwd': 'iigbddauwj'}
批量插入多条记录...
显示所有记录...
{'_id': ObjectId('586243795cd071f570ed3b39'), 'name': 'vvtj', 'passwd': 'iigbddauwj'}
{'_id': ObjectId('5862437a5cd071f570ed3b3a'), 'name': 'nh', 'passwd': 'upyufzknzgdc'}
{'_id': ObjectId('5862437a5cd071f570ed3b3b'), 'name': 'rgf', 'passwd': 'iqdlyjhztq'}
{'_id': ObjectId('5862437a5cd071f570ed3b3c'), 'name': 'dh', 'passwd': 'rgupzruqb'}
{'_id': ObjectId('586243e45cd071f570ed3b3e'), 'name': 'hcq', 'passwd': 'chiwwvxs'}
{'_id': ObjectId('586243e45cd071f570ed3b3f'), 'name': 'yrp', 'passwd': 'kiocdmeerneb'}
{'_id': ObjectId('586243e45cd071f570ed3b40'), 'name': 'hu', 'passwd': 'pknqgfnm'}
{'_id': ObjectId('5862440d5cd071f570ed3b43'), 'name': 'tlh', 'passwd': 'cikouuladgqn'}
{'_id': ObjectId('5862440d5cd071f570ed3b44'), 'name': 'qxf', 'passwd': 'jlsealrqeeel'}
{'_id': ObjectId('5862440d5cd071f570ed3b45'), 'name': 'vlzp', 'passwd': 'wolypmej'}
{'_id': ObjectId('58632e6c5cd07155543cc27a'), 'sid': 2, 'name': 'sgu', 'passwd': 'ogzvdq'}
{'_id': ObjectId('58632e6c5cd07155543cc27b'), 'sid': 3, 'name': 'jiyl', 'passwd': 'atgmhmxr'}
```

```
{'_id':ObjectId('58632e6c5cd07155543cc27c'), 'sid': 4, 'name': 'dbb', 'passwd':
'wmwoeua'}
{'_id': ObjectId('5863305b5cd07155543cc27d'), 'sid': 27, 'name': 'langchao',
'passwd': '123123'}
{'_id': ObjectId('5863305b5cd07155543cc27e'), 'sid': 28, 'name': 'oxp',
'passwd': 'acgjph'}
{'_id': ObjectId('5863305b5cd07155543cc27f'), 'sid': 29, 'name': 'sukj',
'passwd': 'hjtcjf'}
{'_id':ObjectId('5863305b5cd07155543cc280'), 'sid': 30, 'name': 'bf', 'passwd':
'cqerluvk'}
{'_id': ObjectId('5988087533fda81adc0d332f'), 'name': 'hg', 'passwd':
'gmflqxfaxxnv'}
{'_id': ObjectId('5988087533fda81adc0d3330'), 'name': 'ojb', 'passwd':
'rgxodvkprm'}
{'_id': ObjectId('5988087533fda81adc0d3331'), 'name': 'gtdj', 'passwd':
'zigavkysc'}
{'_id': ObjectId('5988087533fda81adc0d3332'), 'name': 'smgt', 'passwd':
'sizvlhdll'}
{'_id': ObjectId('5a33c1cb33fda859b82399d0'), 'name': 'dbu', 'passwd':
'ypdxtqjjafsm'}
{'_id': ObjectId('5a33c1cb33fda859b82399d1'), 'name': 'qg', 'passwd':
'frnoypez'}
{'_id': ObjectId('5a33c1cb33fda859b82399d2'), 'name': 'ky', 'passwd':
'jvzjtcfs'}
{'_id': ObjectId('5a33c1cb33fda859b82399d3'), 'name': 'glnt', 'passwd':
'ejrerztki'}
更新一条记录...
请输入记录的 name：
```

2.5 使用 ORM(对象关系映射)操作数据库

ORM 是对象关系映射(Object Relational Mapping，ORM)的简称，用于实现面向对象编程语言中不同类型系统数据之间的转换。从实现效果上看，ORM 其实是创建了一个可以在编程语言里使用的"虚拟对象数据库"。从另一角度来看，在面向对象编程语言中使用的是对象，而在对象中的数据需要保存到数据库中，或数据库中的数据用来构造对象。在从数据库中提取数据并构造对象或将对象

扫码观看本节视频讲解

数据存入数据库的过程中，有很多代码是可以重复使用的，如果这些重复的功能完全自己实现那就是"重复造轮子"的低效率工作。在这种情况下就诞生了 ORM，它使得从数据库中提取数据来构造对象或将对象数据保存(持久化)到数据库中实现起来更简单。本节将详细讲解在 Python 语言中使用 ORM 操作数据库的知识。

2.5.1 Python 和 ORM

在现实应用中有很多不同的数据库工具，并且其中的大部分系统都包含 Python 接口，能够使开发者更好地利用它们的功能。但是这些不同数据库工具系统的唯一缺点是需要了

解 SQL 语言。如果你是一个更愿意使用 Python 对象而不是 SQL 查询的程序员，并且仍然希望使用关系数据库作为程序的数据后端，那么可能会更加倾向于使用 ORM。

这些 ORM 系统的创始人将纯 SQL 语句进行了抽象化处理，将其实现为 Python 中的对象，这样开发者只要操作这些对象就能完成与生成 SQL 语句相同的任务。一些软件系统也允许一定的灵活性，可以让我们执行几行 SQL 语句。但是大多数情况下，都应该避免使用普通的 SQL 语句。

在 ORM 系统中，数据库表被转化为 Python 类，其中的数据列作为属性，而数据库操作则会作为方法。读者应该会发现，让应用支持 ORM 与使用标准数据库适配器有些相似。由于 ORM 需要执行很多工作，所以一些事情变得更加复杂，或者需要比直接使用适配器更多的代码行。不过，值得欣慰的是，这一点额外工作可以获得更高的开发效率。

在开发过程中，最著名的 Python ORM 是 SQLAlchemy(http://www.qlalchemy.org)和 SQLObject(http://sqlobject.org)。另外一些常用的 Python ORM 还包括 Storm、PyDO/PyDO2、PDO、Dejavu、Durus、QLime 和 ForgetSQL。基于 Web 的大型系统也会包含它们自己的 ORM 组件，如 WebWare MiddleKit 和 Django 的数据库 API。读者需要注意的是，并不是所有的 ORM 都适合于你的应用程序，读者需要根据自己的实际需要来选择。

2.5.2 使用 SQLAlchemy

在 Python 程序中，SQLAlchemy 是一种经典的 ORM。在使用之前需要先安装 SQLAlchemy，安装命令如下：

```
easy_install SQLAlchemy
```

安装成功后的效果如图 2-20 所示。

图 2-20　安装 SQLAlchemy 成功后的效果

例如，在下面的实例文件 SQLAlchemy.py 中，演示了在 Python 程序中使用 SQLAlchemy 操作两种数据库的过程。

源码路径：**codes\2\2-5\SQLAlchemy.py**

```
from distutils.log import import warn as printf
from os.path import dirname
```

```python
from random import randrange as rand
from sqlalchemy import Column, Integer, String, create_engine, exc, orm
from sqlalchemy.ext.declarative import declarative_base
from db import DBNAME, NAMELEN, randName, FIELDS, tformat, cformat, setup
DSNs = {
    'mysql': 'mysql://root@localhost/%s' % DBNAME,
    'sqlite': 'sqlite:///:memory:',
}
Base = declarative_base()
class Users(Base):
    __tablename__ = 'users'
    login  = Column(String(NAMELEN))
    userid = Column(Integer, primary_key=True)
    projid = Column(Integer)
    def __str__(self):
        return ''.join(map(tformat,
            (self.login, self.userid, self.projid)))
class SQLAlchemyTest(object):
    def __init__(self, dsn):
        try:
            eng = create_engine(dsn)
        except ImportError:
            raise RuntimeError()
        try:
            eng.connect()
        except exc.OperationalError:
            eng = create_engine(dirname(dsn))
            eng.execute('CREATE DATABASE %s' % DBNAME).close()
            eng = create_engine(dsn)
        Session = orm.sessionmaker(bind=eng)
        self.ses = Session()
        self.users = Users.__table__
        self.eng = self.users.metadata.bind = eng
    def insert(self):
        self.ses.add_all(
            Users(login=who, userid=userid, projid=rand(1,5)) \
                for who, userid in randName()
        )
        self.ses.commit()
    def update(self):
        fr = rand(1,5)
        to = rand(1,5)
        i = -1
        users = self.ses.query(
            Users).filter_by(projid=fr).all()
        for i, user in enumerate(users):
            user.projid = to
        self.ses.commit()
        return fr, to, i+1
    def delete(self):
        rm = rand(1,5)
        i = -1
        users = self.ses.query(
            Users).filter_by(projid=rm).all()
        for i, user in enumerate(users):
```

```
                self.ses.delete(user)
            self.ses.commit()
            return rm, i+1
        def dbDump(self):
            printf('\n%s' % ''.join(map(cformat, FIELDS)))
            users = self.ses.query(Users).all()
            for user in users:
                printf(user)
            self.ses.commit()
        def __getattr__(self, attr):    # use for drop/create
            return getattr(self.users, attr)
        def finish(self):
            self.ses.connection().close()
def main():
    printf('*** Connect to %r database' % DBNAME)
    db = setup()
    if db not in DSNs:
        printf('\nERROR: %r not supported, exit' % db)
        return
    try:
        orm = SQLAlchemyTest(DSNs[db])
    except RuntimeError:
        printf('\nERROR: %r not supported, exit' % db)
        return
    printf('\n*** Create users table (drop old one if appl.)')
    orm.drop(checkfirst=True)
    orm.create()
    printf('\n*** Insert names into table')
    orm.insert()
    orm.dbDump()
    printf('\n*** Move users to a random group')
    fr, to, num = orm.update()
    printf('\t(%d users moved) from (%d) to (%d)' % (num, fr, to))
    orm.dbDump()
    printf('\n*** Randomly delete group')
    rm, num = orm.delete()
    printf('\t(group #%d; %d users removed)' % (rm, num))
    orm.dbDump()
    printf('\n*** Drop users table')
    orm.drop()
    printf('\n*** Close cxns')
    orm.finish()
if __name__ == '__main__':
    main()
```

► 在上述实例代码中，首先导入了 Python 标准库中的模块(distutils、os.path、random)，然后是第三方或外部模块(sqlalchemy)，最后是应用的本地模块(db)，该模块会提供主要的常量和工具函数。

► 使用了 SQLAlchemy 的声明层，在使用前必须先导入 sqlalchemy.ext.declarative.declarative_base，然后使用它创建一个 Base 类，最后让你的数据子类继承自这个 Base 类。类定义的下一个部分包含一个 __tablename__ 属性，它定义了映射的数据库表名。也可以显式地定义一个低级别的 sqlalchemy.Table 对象，在这种情况下需要将其写为__table__。在大多数情况下使用对象进行数据行的访问，不过也会使用表级别的行为(创建和删除)保存表。接下来是"列"属性，可以通过查阅文档来获取所有可用的数据类型。最后，有一个__str( )__方法定义，用来返回易于阅读的数据行的字符串格式。因为该输出是定制化的(通过 tformat( )函数的协助)，所以不推荐在开发过程中这样使用。

- 通过自定义函数分别实现行的插入、更新和删除操作。插入使用了 session.add_all( )方法，这将使用迭代的方式产生一系列的插入操作。最后，还可以决定是像我们一样进行提交还是进行回滚。update( )和 delete( )方法都存在会话查询的功能，它们使用 query.filter_by( )方法进行查找。随机更新会选择一个成员，通过改变 ID 的方法，将其从一个项目组移动到另一个项目组。计数器会记录有多少用户会受到影响。删除操作则是根据 ID 随机选择一个项目并假设已将其取消，因此项目中的所有员工都将被删除。当要执行操作时，需要通过会话对象进行提交。

- 函数 dbDump( )用于向屏幕上显示输出。该方法从数据库中获取数据行，并按照 db.py 中相似的样式输出数据。

本实例执行后会输出：

```
Choose a database system:

(M)ySQL
(G)adfly
(S)SQLite

Enter choice: S

*** Create users table (drop old one if appl.)

*** Insert names into table

LOGIN      USERID     PROJID
Faye       6812       4
Serena     7003       1
Amy        7209       2
Dave       7306       3
Larry      7311       3
Mona       7404       3
Ernie      7410       3
Jim        7512       3
Angela     7603       3
Stan       7607       3
Jennifer   7608       1
Pat        7711       1
Leslie     7808       4
Davina     7902       4
Elliot     7911       1
Jess       7912       4
Aaron      8312       3
Melissa    8602       4
```

```
*** Move users to a random group
    (1 users moved) from (2) to (1)

LOGIN       USERID      PROJID
Faye        6812        4
Serena      7003        1
Amy         7209        1
Dave        7306        3
Larry       7311        3
Mona        7404        3
Ernie       7410        3
Jim         7512        3
Angela      7603        3
Stan        7607        3
Jennifer    7608        1
Pat         7711        1
Leslie      7808        4
Davina      7902        4
Elliot      7911        1
Jess        7912        4
Aaron       8312        3
Melissa     8602        4

*** Randomly delete group
    (group #1; 5 users removed)

LOGIN       USERID      PROJID
Faye        6812        4
Dave        7306        3
Larry       7311        3
Mona        7404        3
Ernie       7410        3
Jim         7512        3
Angela      7603        3
Stan        7607        3
Leslie      7808        4
Davina      7902        4
Jess        7912        4
Aaron       8312        3
Melissa     8602        4

*** Drop users table

*** Close cxns
```

2.5.3 使用 mongoengine

在 Python 程序中，MongoDB 数据库的 ORM 框架是 mongoengine。在使用 mongoengine 框架之前需要先安装 mongoengine，具体安装命令如下：

```
easy_install mongoengine
```

安装成功后的界面效果如图 2-21 所示。

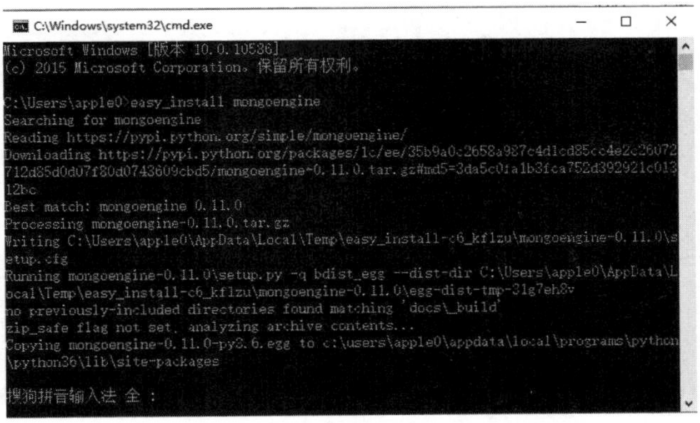

图 2-21 安装 mongoengine 成功后的效果

在运行上述命令之前，必须先确保使用以下命令安装了 pymongo 框架，在本章前面已经安装了 pymongo 框架。

```
easy_install pymongo
```

例如，在下面的实例文件 orm.py 中，演示了在 Python 程序中使用 mongoengine 操作数据库数据的过程。

源码路径：codes\2\2-5\orm.py

```python
import random                            #导入内置模块
from mongoengine import *
connect('test')                          #连接数据库对象'test'
class Stu(Document):                     #定义 ORM 框架类 Stu
    sid = SequenceField()                #"序号"属性表示用户 id
    name = StringField()                 #"用户名"属性
    passwd = StringField()               #"密码"属性
    def introduce(self):                 #定义函数 introduce()显示自己的介绍信息
        print('序号:',self.sid,end=" ")  #打印显示 id
        print('姓名:',self.name,end=' ') #打印显示姓名
        print('密码:',self.passwd)       #打印显示密码
    def set_pw(self,pw):                 #定义函数 set_pw()用于修改密码
        if pw:
            self.passwd = pw             #修改密码
            self.save()                  #保存修改的密码
...省略部分代码...
if __name__ == '__main__':
    print('插入一个文档:')
    stu = Stu(name='langchao',passwd='123123')#创建文档类对象实例 stu，设置用户名和密码
    stu.save()                           #持久化保存文档
    stu = Stu.objects(name='lilei').first()    #查询数据并对类进行初始化
    if stu:
        stu.introduce()                  #显示文档信息
    print('插入多个文档')                #打印提示信息
    for i in range(3):                   #遍历操作
        Stu(name=get_str(2,4),passwd=get_str(6,8)).save()  #插入 3 个文档
```

```
    stus = Stu.objects()                          #文档类对象实例stu
    for stu in stus:                              #遍历所有的文档信息
        stu.introduce()                           #显示所有的遍历文档
    print('修改一个文档')                          #打印提示信息
    stu = Stu.objects(name='langchao').first()    #查询某个要操作的文档
    if stu:
        stu.name='daxie'                          #修改用户名属性
        stu.save()                                #保存修改
        stu.set_pw('bbbbbbbb')                    #修改密码属性
        stu.introduce()                           #显示修改后结果
    print('删除一个文档')                          #打印提示信息
    stu = Stu.objects(name='daxie').first()       #查询某个要操作的文档
    stu.delete()                                  #删除这个文档
    stus = Stu.objects()
    for stu in stus:                              #遍历所有的文档
        stu.introduce()                           #显示删除后结果
```

在上述实例代码中，在导入 mongoengine 库和连接 MongoDB 数据库后，定义了一个继承于类 Document 的子类 Stu。在主程序中通过创建类的实例，并调用其方法 save( )将类固化到数据库；通过类 Stu 中的方法 objects( )来查询数据库并映射为类 Stu 的实例，并调用其自定义方法 introduce( )来显示载入的信息。然后插入 3 个文档信息，并调用方法 save( )固化到数据库，通过调用类中的自定义方法 set_pw( )修改数据并存入数据库。最后通过调用类中的方法 delete( )从数据库中删除一个文档。

开始测试程序，在运行本实例程序时，必须在 CMD 控制台中启动 MongoDB 服务，并且使上述控制台界面处于打开状态。下面是开启 MongoDB 服务的命令：

```
mongod --dbpath "h:\data"
```

在上述命令中，"h:\data"是保存 MongoDB 数据库数据的目录。

本实例执行后的效果如图 2-22 所示。

图 2-22　执行后的效果

第 3 章

绘制散点图和折线图

在现实应用中,经常使用统计图来展示数据分析的结果,这样可以更加直观地展示数据分析结果。常用的统计图类型有折线图、散点图、柱状图、饼状图等。本章将详细讲解使用 Python 绘制散点图和折线图的知识。

3.1 绘制散点图

在数据可视化分析工作中，经常需要绘制与数据统计相关的图形，如折线图、散点图、直方图等。本节将详细讲解在 Python 程序中使用各种库绘制散点图的知识。

3.1.1 绘制一个简单的点

扫码观看本节视频讲解

假设你有一堆的数据样本，想要找出其中的异常值，那么最直观的方法就是将它们画成散点图。最简单的散点图只有一个点。例如，在下面的实例文件 dian01.py 中，演示了使用 Matplotlib 绘制只有两个点的散点图的过程。

> 源码路径：codes\3\3-1\dian01.py

```
import matplotlib.pyplot as plt        #导入pyplot包,并缩写为plt
#定义2个点的x集合和y集合
x=[1,2]
y=[2,4]
plt.scatter(x,y)                       #绘制散点图
plt.show()                             #展示绘画框
```

在上述实例代码中绘制了拥有两个点的散点图，向函数 scatter( ) 传递了两个分别包含 x 值和 y 值的列表。执行效果如图 3-1 所示。

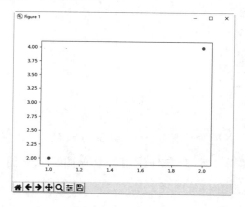

图 3-1 执行效果

在上述实例中，可以进一步调整坐标轴的样式，如可以加上以下代码：

```
#[]里的4个参数分别表示X轴起始点,X轴结束点,Y轴起始点,Y轴结束点
plt.axis([0,10,0,10])
```

3.1.2 添加标题和标签

可以设置散点图的输出样式，如添加标题、给坐标轴添加标签，并确保所有文本都能

看清。请看下面的实例文件 dian02.py，功能是使用 Matplotlib 函数 scatter( )绘制一系列点，然后设置显示的标题和标签。

源码路径：codes\3\3-1\dian02.py

```python
"""使用scatter()绘制散点图"""
import matplotlib.pyplot as plt

x_values = range(1, 6)
y_values = [x*x for x in x_values]
'''
scatter()
x:横坐标 y:纵坐标 s:点的尺寸
'''
plt.scatter(x_values, y_values, s=50)

# 设置图表标题并给坐标轴加上标签
plt.title('Square Numbers', fontsize=24)
plt.xlabel('Value', fontsize=14)
plt.ylabel('Square of Value', fontsize=14)

# 设置刻度标记的大小
plt.tick_params(axis='both', which='major', labelsize=14)
plt.show()
```

执行效果如图 3-2 所示。

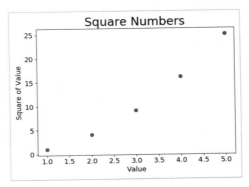

图 3-2　执行效果

3.1.3　绘制 10 个点

请看下面的实例文件 dian03.py，功能是使用 Matplotlib 函数 scatter( )绘制 10 个点。

源码路径：codes\3\3-1\dian03.py

```python
import matplotlib.pyplot as plt
import numpy as np
# 保证图片在浏览器内正常显示
N = 10                                      # 10个点
x = np.random.rand(N)
y = np.random.rand(N)
```

```
plt.scatter(x, y)
plt.show()
```

执行效果如图 3-3 所示。

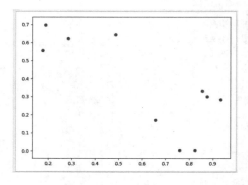

图 3-3　执行效果

3.1.4　修改散点的大小

请看下面的实例文件 dian04.py，功能是使用 Matplotlib 函数 scatter( )绘制 10 个点并随机设置点的大小。

源码路径：codes\3\3-1\dian04.py

```
import matplotlib.pyplot as plt
import numpy as np

N = 10                                    # 10个点
x = np.random.rand(N)
y = np.random.rand(N)

s = (30*np.random.rand(N))**2             # 每个点随机大小
plt.scatter(x, y, s=s)
plt.show()
```

执行效果如图 3-4 所示。

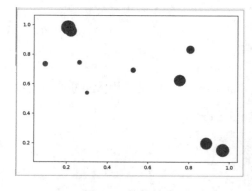

图 3-4　执行效果

3.1.5 设置散点的颜色和透明度

请看下面的实例文件 dian05.py,功能是使用 Matplotlib 函数 scatter( )绘制散点图,分别设置散点的颜色和透明度。

源码路径:codes\3\3-1\dian05.py

```python
import matplotlib.pyplot as plt
import numpy as np

N = 10                              # 10 个点
x = np.random.rand(N)
y = np.random.rand(N)
s = (30*np.random.rand(N))**2       # 每个点随机大小

c = np.random.rand(N)               # 随机颜色
plt.scatter(x, y, s=s, c=c, alpha=0.5)
plt.show()
```

执行效果如图 3-5 所示。

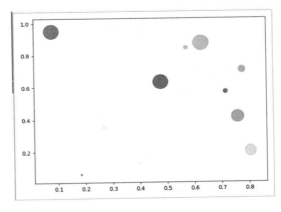

图 3-5 执行效果

3.1.6 修改散点的形状

请看下面的实例文件 dian06.py,功能是使用 Matplotlib 函数 scatter( )绘制散点图,然后使用函数 scatter( )的属性 marker 设置散点的形状。

源码路径:codes\3\3-1\dian06.py

```python
import matplotlib.pyplot as plt
import numpy as np

N = 10                              # 10 个点
x = np.random.rand(N)
y = np.random.rand(N)
s = (30*np.random.rand(N))**2
c = np.random.rand(N)
```

```
plt.scatter(x, y, s=s, c=c, marker='^', alpha=0.5)
plt.show()
```

执行效果如图 3-6 所示。

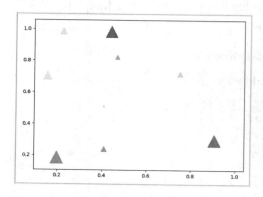

图 3-6 执行效果

3.1.7 绘制两组数据的散点图

请看下面的实例文件 dian07.py，功能是使用 Matplotlib 函数 scatter( )绘制两组数据的散点图，然后使用函数 scatter( )的属性 marker 设置这两组散点图的形状。

源码路径：codes\3\3-1\dian07.py

```
import matplotlib.pyplot as plt
import numpy as np
N = 10                                  # 10 个点
x1 = np.random.rand(N)
y1 = np.random.rand(N)
x2 = np.random.rand(N)
y2 = np.random.rand(N)
plt.scatter(x1, y1, marker='o')
plt.scatter(x2, y2, marker='^')
plt.show()
```

执行效果如图 3-7 所示。

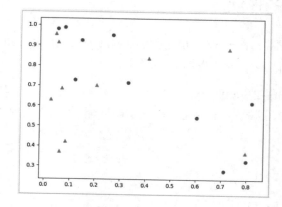

图 3-7 执行效果

3.1.8 为散点图设置图例

图例是集中于地图一角或一侧的地图上各种符号和颜色所代表内容与指标的说明，有助于更好地认识地图。请看下面的实例文件 dian08.py。首先使用 Matplotlib 函数 scatter( ) 绘制两组数据的散点图；然后使用函数 scatter( )的属性 marker 设置这两组散点图的形状；最后用函数 scatter( )的属性 label 设置了这两组散点图的图例。

源码路径：codes\3\3-1\dian08.py

```
import matplotlib.pyplot as plt
import numpy as np

N = 10                    # 10个点
x1 = np.random.rand(N)
y1 = np.random.rand(N)
x2 = np.random.rand(N)
y2 = np.random.rand(N)
plt.scatter(x1, y1, marker='o', label="circle")
plt.scatter(x2, y2, marker='^', label="triangle")
plt.legend(loc='best')
plt.show()
```

执行效果如图 3-8 所示。

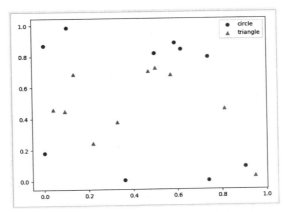

图 3-8　执行效果

3.1.9 自定义散点图样式

在实际应用中，经常需要绘制散点图并设置各个数据点的样式。例如，可能想以一种颜色显示较小的值，而用另一种颜色显示较大的值。当绘制大型数据集时，还需要对每个点都设置同样的样式，再使用不同的样式选项重新绘制某些点，这样可以突出显示它们的效果。在 Matplotlib 库中，可以使用函数 scatter( )绘制单个点，通过传递 x 点和 y 点坐标的方式在指定位置绘制一个点。例如，在下面的实例文件 dian09.py 中，演示了使用 matplotlib 绘制指定样式散点图的过程。

源码路径：codes\3\3-1\dian09.py

```python
import matplotlib.pyplot as plt
from pylab import *
mpl.rcParams['font.sans-serif'] = ['SimHei'] #指定默认字体

mpl.rcParams['axes.unicode_minus'] = False #解决保存图像是负号'-'显示为方块的问题
x_values = list(range(1, 1001))
y_values = [x**2 for x in x_values]

plt.scatter(x_values, y_values, c=(0, 0, 0.8), edgecolor='red', s=40)

# 设置图表标题，并设置坐标轴标签
plt.title("大中华区销售统计表", fontsize=24)
plt.xlabel("节点", fontsize=14)
plt.ylabel("销售数据", fontsize=14)

# 设置刻度大小
plt.tick_params(axis='both', which='major', labelsize=14)

# 设置每个坐标轴的取值范围
plt.axis([0, 110, 0, 1100])

plt.show()
```

(1) 第2~4行代码：导入字体库，设置中文字体，并解决负号"-"显示为方块的问题。

(2) 第5行和第6行代码：使用 Python 循环实现自动计算数据功能，首先创建一个包含 x 值的列表，其中包含数字 1~1001。接下来创建一个生成 y 值的列表解析，它能够遍历 x 值(for x in x_values)，计算其平方值(x**2)，并将结果存储到列表 y_values 中。

(3) 第7行代码：将输入列表和输出列表传递给函数 scatter( )。另外，因为 matplotlib 允许给散列点图中的各个点设置一个颜色，默认为蓝色点和黑色轮廓。所以，当在散列点图中包含的数据点不多时效果会很好。但是当需要绘制很多个点时，这些黑色的轮廓可能会黏连在一起，此时需要删除数据点的轮廓。所以在本行代码中，在调用函数 scatter( )时传递了实参：edgecolor='none'。为了修改数据点的颜色，在此向函数 scatter( )传递参数 c，并将其设置为要使用的颜色的名称 red。

> **注 意**
>
> 颜色映射(Colormap)是一系列颜色，它们从起始颜色渐变到结束颜色。在可视化视图模型中，颜色映射用于突出数据的规律，如可能需要用较浅的颜色来显示较小的值，并使用较深的颜色来显示较大的值。在模块 pyplot 中内置一组颜色映射，要想使用这些颜色映射，需要告诉 pyplot 应该如何设置数据集中每个点的颜色。

(4) 第15行代码：因为这个数据集较大，所以将点设置得较小，在本行代码中使用函数 axis( )指定每个坐标轴的取值范围。函数 axis( )要求提供四个值，即 x 和 y 坐标轴的最小值和最大值。此处将 x 坐标轴的取值范围设置为 0~110，并将 y 坐标轴的取值范围设置为 0~1100。

(5) 第 16 行(最后一行)代码：使用函数 plt.show( )显示绘制的图形。当然也可以让程序自动将图表保存到一个文件中，此时只需对 plt.show( )函数的调用替换为对 plt.savefig( )函数的调用即可。

```
plt.savefig (' plot.png' , bbox_inches='tight' )
```

在上述代码中，第 1 个实参用于指定要以什么样的文件名保存图表，这个文件将存储到当前实例文件 dianyang.py 所在的目录中。第 2 个实参用于指定将图表多余的空白区域裁剪掉。如果要保留图表周围多余的空白区域，可省略这个实参。

执行效果如图 3-9 所示。

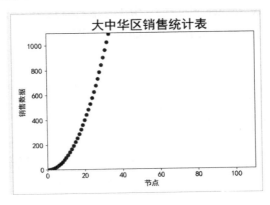

图 3-9　执行效果

3.1.10　使用 pygal 绘制散点图

在 Python 程序中，可以使用库 pygal 实现数据的可视化操作功能，生成 SVG 格式的图形文件。SVG(Scalable Vector Graphics，可缩放矢量图形)是一种常见的数据分析文件格式，是一种矢量图格式，可以使用浏览器打开 SVG 文件，可以方便地与之交互。对于需要在尺寸不同的屏幕上显示的图表，SVG 会变得很有用，可以自动缩放，自适应观看者的屏幕。例如，在下面的实例文件 dian10.py 中，演示了使用 pygal 绘制指定样式散点图的过程。

源码路径：codes\3\3-1\dian10.py

```
import random

import pygal

scatter_plot_chart = pygal.XY(stroke=False)
scatter_plot_chart.title = "散点图"
list_a = []
list_b = []
list_c = []
for i in range(20):
    a = random.uniform(0,5)
    b = random.uniform(0,5)
    list_a.append((a,b))
```

```
for j in range(20):
    a = random.uniform(0,5)
    b = random.uniform(0,5)
    list_b.append((a,b))
for s in range(20):
    a = random.uniform(0,5)
    b = random.uniform(0,5)
    list_c.append((a,b))
scatter_plot_chart.add('A', list_a)
scatter_plot_chart.add('B', list_b)
scatter_plot_chart.add('C', list_c)
scatter_plot_chart.render_to_file('xy_scatter_plot.svg')
```

执行后会生成文件 xy_scatter_plot.svg，用浏览器打开这个文件后会看到绘制的散点图，将光标放在某个散点上时会悬浮显示这个点的值，如图 3-10 所示。

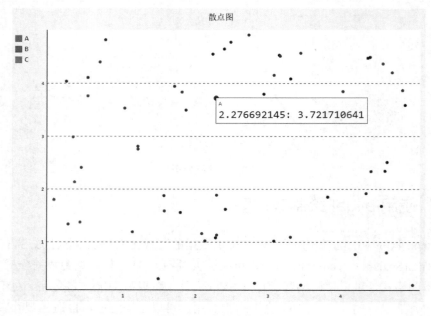

图 3-10　绘制的散点图

3.2　绘制折线图

除了前面介绍的散点图外，在数据可视化工作中还经常需要绘制折线图。本节将详细讲解在 Python 程序中使用各种库绘制折线图的知识。

3.2.1　绘制最简单的折线

扫码观看本节视频讲解

可以使用库 Matplotlib 绘制折线图。例如，在下面的实例文件 zhe01.py 中，使用 matplotlib 绘制一个简单的折线图，现实样式是默认的效果。

源码路径：daima\3\3-2\zhe01.py

```
import matplotlib.pyplot as plt
squares = [1, 4, 9, 16, 25]
plt.plot(squares)
plt.show()
```

在上述实例代码中，使用平方数序列 1、4、9、16 和 25 绘制一个折线图，在具体实现时，只需向 matplotlib 提供这些平方数序列数字就能完成绘制工作。

(1) 导入模块 pyplot，并给它指定别名 plt，以免反复输入 pyplot，在模块 pyplot 中包含很多用于生成图表的函数。

(2) 创建一个列表，在其中存储前述平方数。

(3) 将创建的列表传递给函数 plot( )，该函数会根据这些数字绘制出有意义的图形。

(4) 通过函数 plt.show( )打开 matplotlib 查看器，并显示绘制的图形。

执行效果如图 3-11 所示。

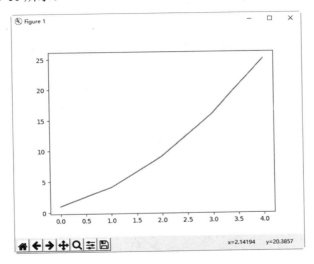

图 3-11　执行效果

3.2.2　设置标签文字和线条粗细

本章前面实例文件 zhe01.py 的执行效果不够完美，开发者可以对绘制的线条样式进行灵活设置，如可以设置线条粗细、实现数据准确性校正等操作。例如，在下面的实例文件 zhe02.py 中，演示了使用 matplotlib 绘制指定样式折线图效果的过程。

源码路径：daima\3\3-2\zhe02.py

```
import matplotlib.pyplot as plt                    #导入模块
input_values = [1, 2, 3, 4, 5]
squares = [1, 4, 9, 16, 25]
plt.plot(input_values, squares, linewidth=5)
# 设置图表标题，并在坐标轴上添加标签
plt.title("Numbers", fontsize=24)
```

```
plt.xlabel("Value", fontsize=14)
plt.ylabel("ARG Value", fontsize=14)
# 设置单位刻度的大小
plt.tick_params(axis='both', labelsize=14)
plt.show()
```

(1) 第 4 行代码中的 linewidth=5：设置线条的粗细。

(2) 第 4 行代码中的函数 plot( )：当向函数 plot( )提供一系列数字时，它会假设第一个数据点对应的 x 坐标值为 0，但是实际上第一个点对应的 x 值为 1。为改变这种默认行为，可以给函数 plot( )同时提供输入值和输出值，这样使用函数 plot( )可以正确地绘制数据，因为同时提供了输入值和输出值，而无须对输出值的生成方式进行假设，所以最终绘制出的图形是正确的。

(3) 第 6 行代码中的函数 title( )：设置图表的标题。

(4) 第 6～8 行中的参数 fontsize：设置图表中的文字大小。

(5) 第 7 行中的函数 xlabel( )和第 8 行中的函数 ylabel( )：分别设置 x 轴的标题和 y 轴的标题。

(6) 第 10 行中的函数 tick_params( )：设置刻度样式，其中指定的实参将影响 x 轴和 y 轴上的刻度(axis='both')，并将刻度标记的字体大小设置为 14(labelsize=14)。

执行效果如图 3-12 所示。

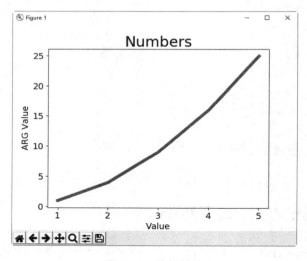

图 3-12　执行效果

3.2.3　绘制 1000 个点组成折线图

可以用很多个点组成一个折线，请看下面的实例文件 zhe03.py，功能是使用 matplotlib 函数 scatter( )绘制 1000 个散点组成折线图。

 源码路径：**daima\3\3-2\zhe03.py**

```
import matplotlib.pyplot as plt
```

```
x_values = range(1, 1001)
y_values = [x*x for x in x_values]
'''
scatter()
x:横坐标 y:纵坐标 s:点的尺寸
'''
plt.scatter(x_values, y_values, s=10)

# 设置图表标题并给坐标轴加上标签
plt.title('Square Numbers', fontsize=24)
plt.xlabel('Value', fontsize=14)
plt.ylabel('Square of Value', fontsize=14)

# 设置刻度标记的大小
plt.tick_params(axis='both', which='major', labelsize=14)

# 设置每个坐标轴的取值范围
plt.axis([0, 1100, 0, 1100000])
plt.show()
```

在上述代码中，函数 axis( ) 有 4 个参数值[xmin, xmax, ymin, ymax]，分别表示 x、y 坐标轴的最小值和最大值。执行效果如图 3-13 所示。

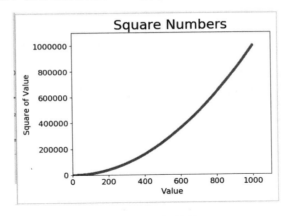

图 3-13　执行效果

3.2.4　绘制渐变色的折线图

使用颜色映射可以绘制渐变色的折线图，颜色映射是一系列颜色，它们从起始颜色渐变到结束颜色。在可视化中，颜色映射用于突出数据的规律。例如，可以用较浅的颜色来显示较小的值，并使用较深的颜色来显示较大的值。请看下面的实例文件 zhe04.py，功能是使用 matplotlib 函数 scatter( ) 绘制渐变色的折线图。

源码路径：daima\3\3-2\zhe04.py

```
import matplotlib.pyplot as plt

x_values = range(1, 1001)
y_values = [x*x for x in x_values]
```

```
'''
scatter()
x:横坐标 y:纵坐标 s:点的尺寸
'''
plt.scatter(x_values, y_values, c=y_values, cmap=plt.cm.Blues,
    edgecolors='none', s=10)

# 设置图表标题并给坐标轴加上标签
plt.title('Square Numbers', fontsize=24)
plt.xlabel('Value', fontsize=14)
plt.ylabel('Square of Value', fontsize=14)

# 设置刻度标记的大小
plt.tick_params(axis='both', which='major', labelsize=14)

# 设置每个坐标轴的取值范围
plt.axis([0, 1100, 0, 1100000])
plt.show()
```

在上述代码中，将参数 c 设置成一个 y 值列表，并使用参数 cmap 告诉 pyplot 使用哪个颜色映射。这些代码将 y 值较小的点显示为浅蓝色，并将 y 值较大的点显示为深蓝色。执行效果如图 3-14 所示。

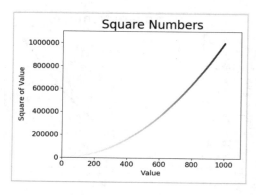

图 3-14　执行效果

3.2.5　绘制多幅子图

在 Matplotlib 绘图系统中，可以显式地控制图像、子图和坐标轴。Matplotlib 中的"图像"指的是用户界面看到的整个窗口内容。在图像里面有所谓"子图"，子图的位置是由坐标网格确定的，而"坐标轴"却不受此限制，可以放在图像的任意位置。当调用 plot( )函数时，Matplotlib 调用 gca( )函数以及 gcf( )函数来获取当前的坐标轴和图像。如果无法获取图像，则会调用 figure( )函数来创建一个图像。从严格意义上来说，是使用 subplot(1,1,1) 创建只有一个子图的图像。

在 Matplotlib 绘图系统中，所谓"图像"就是 GUI 里以 Figure #为标题的那些窗口。图像编号从 1 开始，与 MATLAB 的风格一致，而与 Python 从 0 开始编号的风格不同。表 3-1 中的参数是图像的属性。

表 3-1 图像的属性

参　数	默　认　值	描　述
num	1	图像的数量
figsize	figure.figsize	图像的长和宽(英寸)
dpi	figure.dpi	分辨率(点/英寸)
facecolor	figure.facecolor	绘图区域的背景颜色
edgecolor	figure.edgecolor	绘图区域边缘的颜色
frameon	True	是否绘制图像边缘

例如，在下面的实例文件 zhe05.py 中，演示了让一个折线图和一个散点图出现在同一个绘画框中的过程。

源码路径：codes\3\3-2\zhe05.py

```
import matplotlib.pyplot as plt  #将绘画框进行对象化
fig=plt.figure()  #将p1定义为绘画框的子图,211表示将绘画框划分为2行1列,最后的1表示第一幅图
p1=fig.add_subplot(211)
x=[1,2,3,4,5,6,7,8]
y=[2,1,3,5,2,6,12,7]
p1.plot(x,y) #将p2定义为绘画框的子图,212表示将绘画框划分为2行1列,最后的2表示第二幅图
p2=fig.add_subplot(212)
a=[1,2]
b=[2,4]
p2.scatter(a,b)
plt.show()
```

上述代码执行后的效果如图 3-15 所示。

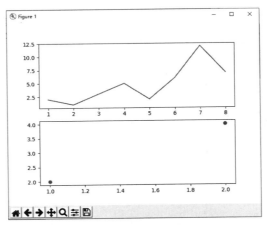

图 3-15　执行后的效果

3.2.6　绘制正弦函数和余弦函数曲线

在 Python 程序中，最简单绘制曲线的方式是使用数学中的正弦函数或余弦函数。例如，在下面的实例文件 zhe06.py 中，演示了使用正弦函数和余弦函数绘制曲线的过程。

源码路径：codes\3\3-2\zhe06.py

```
from pylab import *
X = np.linspace(-np.pi, np.pi, 256,endpoint=True)
C,S = np.cos(X), np.sin(X)
plot(X,C)
plot(X,S)
show()
```

执行后的效果如图 3-16 所示。

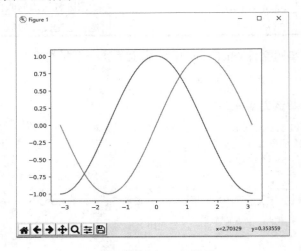

图 3-16　执行后的效果

在上述实例中，展示的是使用 Matplotlib 默认配置的效果。其实开发者可以调整大多数的默认配置，如图片大小和分辨率(dpi)、线宽、颜色、风格、坐标轴以及网格的属性、文字与字体属性等。但是，Matplotlib 的默认配置在大多数情况下已经做得足够好，开发人员可能很少想更改这些默认配置。例如，在下面的实例文件 zhe07.py 中，展示了使用 Matplotlib 的默认配置和自定义绘图样式的过程。

源码路径：codes\3\3-2\zhe07.py

```
# 导入 matplotlib 的所有内容(nympy 可以用 np 名来使用)
from pylab import *
# 创建一个 8 * 6 点(point)的图，并设置分辨率为 80
figure(figsize=(8,6), dpi=80)
# 创建一个新的 1 * 1 的子图，接下来的图样绘制在其中的第 1 块(也是唯一的一块)
subplot(1,1,1)
X = np.linspace(-np.pi, np.pi, 256,endpoint=True)
C,S = np.cos(X), np.sin(X)
# 绘制余弦曲线，使用蓝色的、连续的、宽度为 1 (像素)的线条
plot(X, C, color="blue", linewidth=1.0, linestyle="-")
# 绘制正弦曲线，使用绿色的、连续的、宽度为 1 (像素)的线条
plot(X, S, color="green", linewidth=1.0, linestyle="-")
# 设置横轴的上下限
xlim(-4.0,4.0)
# 设置横轴记号
xticks(np.linspace(-4,4,9,endpoint=True))
```

```
# 设置纵轴的上下限
ylim(-1.0,1.0)
# 设置纵轴记号
yticks(np.linspace(-1,1,5,endpoint=True))
# 以分辨率 72 来保存图片
# savefig("exercice_2.png",dpi=72)
# 在屏幕上显示
show()
```

上述实例代码中的配置与默认配置完全相同，可以在交互模式中修改其中的值来观察效果。执行后的效果如图 3-17 所示。

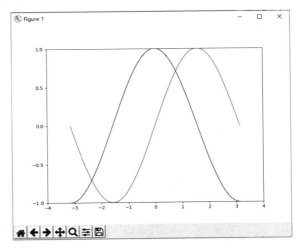

图 3-17　执行后的效果

在绘制曲线时可以改变线条的颜色和粗细，如以蓝色和红色分别表示余弦函数和正弦函数，然后将线条变粗一点，接着在水平方向拉伸一下整个图。

```
...
figure(figsize=(10,6), dpi=80)
plot(X, C, color="blue", linewidth=2.5, linestyle="-")
plot(X, S, color="red", linewidth=2.5, linestyle="-")
...
```

此时的执行效果如图 3-18 所示。

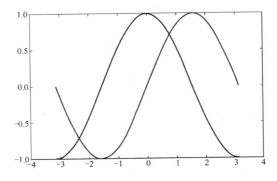

图 3-18　改变线条的颜色和粗细

请看下面的实例文件 zhe07-1.py，功能是使用 Matplotlib 分别绘制正弦曲线和余弦曲线，并分别在绘制的曲线上标注正弦和余弦表达式。

源码路径：codes\3\3-2\zhe07-1.py

```python
import matplotlib.pyplot as plt    # 导入绘图模块
import numpy as np                 # 导入需要生成数据的numpy模块
'''
第一种方式 text()
 text(x,y,s,fontdict=None, withdash=False)
    参数说明：(1)x,y 坐标位置
            (2) 显示的文本
'''
x = np.arange(0, 2 * np.pi, 0.01)
plt.plot(np.sin(x))
'''x,y 代表着坐标系中数值'''
plt.text(20, 0, 'sin(0) = 0')
'''
第二种方式 figtext()
    使用figtext时,x、y代表相对值,图片的宽度
'''
x2 = np.arange(0, 2 * np.pi, 0.01)
plt.plot(np.cos(x2))
''''''
plt.figtext(0.5, 0.5, 'cos(0)=0')
plt.show()
```

执行效果如图 3-19 所示。

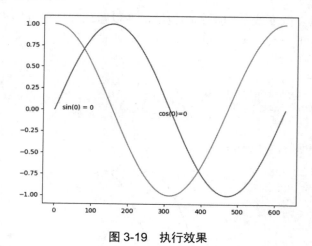

图 3-19 执行效果

3.2.7 绘制 3 条不同的折线

在使用 Matplotlib 绘制折线时，可以使用函数 plot( ) 设置样式，如红色虚线、蓝色正方形和绿色三角形等。请看下面的实例文件 zhe08.py，功能是使用 Matplotlib 绘制 3 条不同的

折线。

源码路径：codes\3\3-2\zhe08.py

```python
import numpy as np
import matplotlib.pyplot as plt

# 间隔200ms 均匀采样
t = np.arange(0., 5., 0.2)

#红色虚线、蓝色正方形和绿色三角形
plt.plot(t, t, 'r--', t, t ** 2, 'bs', t, t ** 3, 'g^')
plt.show()
```

执行后的效果如图 3-20 所示。

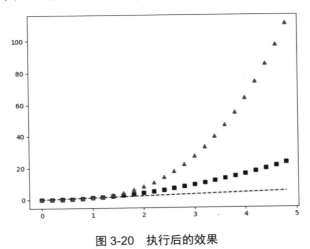

图 3-20　执行后的效果

3.2.8　绘制浏览器市场占有率变化折线图

除了使用 Matplotlib 外，也可以使用库 pygal 绘制折线。请看下面的实例文件 zhe09.py，功能是使用 pygal 绘制在某时间段内浏览器产品的市场占有率变化折线图。

源码路径：codes\3\3-2\zhe09.py

```
import pygal

line_chart = pygal.Line()
line_chart.title = 'Browser usage evolution (in %)'
line_chart.x_labels = map(str, range(2002, 2013))
line_chart.add('Firefox', [None, None,    0, 16.6,   25,   31, 36.4, 45.5, 46.3, 42.8, 37.1])
line_chart.add('Chrome',  [None, None, None, None, None, None,    0,  3.9, 10.8, 23.8, 35.3])
line_chart.add('IE',      [85.8, 84.6, 84.7, 74.5,   66, 58.6, 54.7, 44.8, 36.2, 26.6, 20.1])
line_chart.add('Others',  [14.2, 15.4, 15.3,  8.9,    9, 10.4,  8.9,  5.8,  6.7,  6.8,  7.5])
line_chart.render_to_file('bar_chart.svg')
```

执行效果如图 3-21 所示。

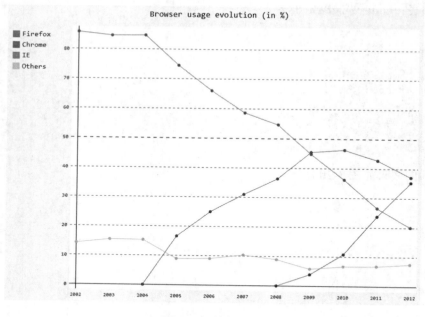

图 3-21　执行效果

3.2.9　绘制 XY 线图

XY 线是将各个点用直线连接起来的折线图，在绘制时需提供一个横纵坐标元组作为元素的列表。使用 pygal 绘制 XY 线图的方法十分简单，只需调用库 pygal 中的 XY( )方法即可。例如，在下面的实例文件 zhe10.py 中，演示了使用 XY( )方法绘制两条 XY 余弦曲线图的过程。

源码路径：codes\3\3-2\zhe10.py

```python
import pygal
from math import cos

xy_chart = pygal.XY()
xy_chart.title = 'XY 余弦曲线图'
xy_chart.add('x = cos(y)', [(cos(x / 10.), x / 10.) for x in range(-50, 50, 5)])
xy_chart.add('y = cos(x)', [(x / 10., cos(x / 10.)) for x in range(-50, 50, 5)])
xy_chart.add('x = 1',  [(1, -5),  (1, 5)])
xy_chart.add('x = -1', [(-1, -5), (-1, 5)])
xy_chart.add('y = 1',  [(-5, 1),  (5, 1)])
xy_chart.add('y = -1', [(-5, -1), (5, -1)])
xy_chart.render_to_file('bar_chart.svg')
```

执行后会创建生成 XY 余弦曲线图文件 bar_chart.svg，打开后的效果如图 3-22 所示。

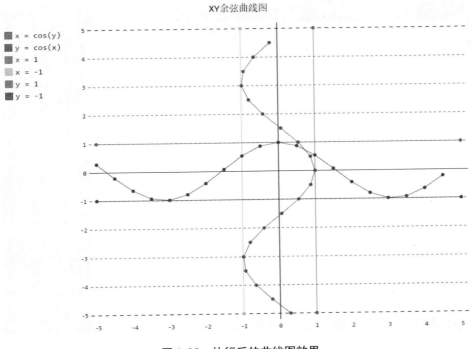

图 3-22　执行后的曲线图效果

3.2.10　绘制水平样式的浏览器市场占有率变化折线图

请看下面的实例文件 zhe11.py，功能是使用 pygal 绘制在某时间段内浏览器产品的市场占有率变化折线图，但这个折线图是水平样式的。

源码路径：**codes\3\3-2\zhe11.py**

```
import pygal

line_chart = pygal.HorizontalLine()
line_chart.title = 'Browser usage evolution (in %)'
line_chart.x_labels = map(str, range(2002, 2013))
line_chart.add('Firefox', [None, None,    0, 16.6,   25,   31, 36.4, 45.5, 46.3, 42.8, 37.1])
line_chart.add('Chrome',  [None, None, None, None, None, None,    0,  3.9, 10.8, 23.8, 35.3])
line_chart.add('IE',      [85.8, 84.6, 84.7, 74.5,   66, 58.6, 54.7, 44.8, 36.2, 26.6, 20.1])
line_chart.add('Others',  [14.2, 15.4, 15.3,  8.9,    9, 10.4,  8.9,  5.8,  6.7,  6.8,  7.5])
line_chart.range = [0, 100]
line_chart.render_to_file('bar_chart.svg')
```

执行后的效果如图 3-23 所示。

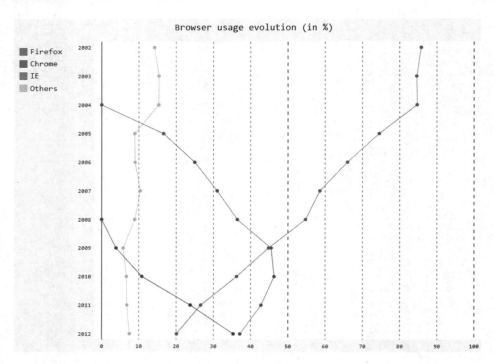

图 3-23 执行后的折线图效果

3.2.11 绘制叠加折线图

请看下面的实例文件 zhe12.py，功能是使用 pygal 绘制浏览器市场占有率的叠加折线图。

源码路径：codes\3\3-2\zhe12.py

```
import pygal

line_chart = pygal.StackedLine(fill=True)
line_chart.title = 'Browser usage evolution (in %)'
line_chart.x_labels = map(str, range(2002, 2013))
line_chart.add('Firefox', [None, None, 0, 16.6,   25, 31, 36.4, 45.5, 46.3, 42.8, 37.1])
line_chart.add('Chrome',  [None, None, None, None, None, None,  0, 3.9, 10.8, 23.8, 35.3])
line_chart.add('IE',     [85.8, 84.6, 84.7, 74.5,   66, 58.6, 54.7, 44.8, 36.2, 26.6, 20.1])
line_chart.add('Others', [14.2, 15.4, 15.3, 8.9,    9, 10.4, 8.9, 5.8, 6.7, 6.8,  7.5])
line_chart.render_to_file('bar_chart.svg')
```

执行后的效果如图 3-24 所示。

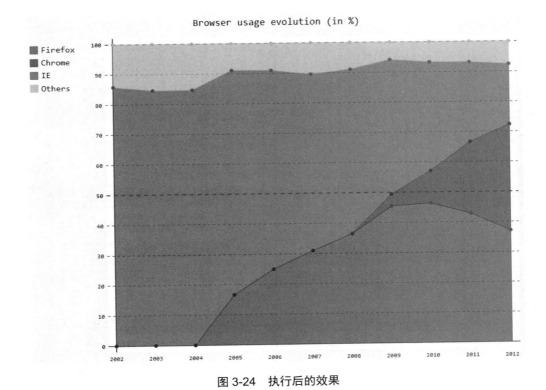

图 3-24 执行后的效果

3.2.12 绘制某网站用户访问量折线图

在使用 pygal 绘制折线图时,可以设置坐标轴标签的显示样式,如可以在 x 轴标签显示"年-月-日"信息。请看下面的实例文件 zhe13.py,功能是使用 pygal 绘制某时间段内某网站用户访问量折线图,x 轴标签显示的是日期信息。

源码路径:codes\3\3-2\zhe13.py

```
from datetime import datetime
import pygal

date_chart = pygal.Line(x_label_rotation=20)
date_chart.x_labels = map(lambda d: d.strftime('%Y-%m-%d'), [
 datetime(2013, 1, 2),
 datetime(2013, 1, 12),
 datetime(2013, 2, 2),
 datetime(2013, 2, 22)])
date_chart.add("Visits", [300, 412, 823, 672])
date_chart.render_to_file('bar_chart.svg')
```

执行后的效果如图 3-25 所示。

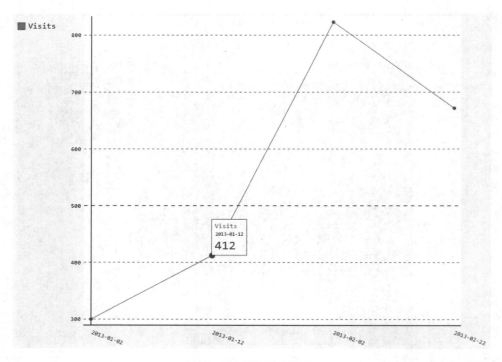

图 3-25　执行后的效果

3.3　绘制其他类型的散点图和折线图

除了在本章前面介绍的散点图和折线图外，还可以使用 Python 绘制功能更加强大的散点图和折线图。本节将详细讲解绘制其他类型散点图和折线图的知识。

3.3.1　绘制随机漫步图

扫码观看本节视频讲解

随机漫步(Random walk)是一种数学统计模型，它由一连串轨迹所组成。其中每一次都是随机的，它能用来表示不规则的变动形式。气体或液体中分子活动的轨迹等可作为随机漫步的模型，如同一个人酒后乱步所形成的随机记录。在 1903 年由卡尔·皮尔逊首次提出随机漫步这一概念，目前已经被广泛应用于生态学、经济学、心理学、计算科学、物理学、化学和生物学等领域，用来说明这些领域内观察到的行为和过程，因而是记录随机活动的基本模型。

1. 在 Python 程序中生成随机漫步数据

在 Python 程序中生成随机漫步数据后，可以使用 Matplotlib 灵活方便地将这些数据展现出来。随机漫步的行走路径很有自己的特色，每次行走动作都完全是随机的，没有任何明确的方向，漫步结果是由一系列随机决策决定的。例如，漂浮在水滴上的花粉因不断受到水分子的挤压而在水面上移动。水滴中的分子运动是随机的，因此花粉在水面上的运动路径犹如随机漫步。

为了在 Python 程序中模拟随机漫步的过程，在下面的实例文件 random_walk.py 中创建一个名为 RandomA 的类，此类可以随机选择前进方向。类 RandomA 需要用到 3 个属性，其中一个是存储随机漫步次数的变量，其他两个是列表，分别用于存储随机漫步经过的每个点的 x 坐标和 y 坐标。

源码路径：codes\3\3-3\random_walk.py

```python
from random import choice
class RandomA():
    """能够随机生成漫步数据的类"""
    def __init__(self, num_points=5100):
        """初始化随机漫步属性"""
        self.num_points = num_points
        # 所有的随机漫步开始于 (0, 0).
        self.x_values = [0]
        self.y_values = [0]
    def shibai(self):
        """计算在随机漫步中包含的所有点"""
        # 继续漫步,直到达到所需长度为止
        while len(self.x_values) < self.num_points:
            # 决定前进的方向,沿着这个方向前进的距离
            x_direction = choice([1, -1])
            x_distance = choice([0, 1, 2, 3, 4])
            x_step = x_direction * x_distance
            y_direction = choice([1, -1])
            y_distance = choice([0, 1, 2, 3, 4])
            y_step = y_direction * y_distance
            # 不能原地踏步
            if x_step == 0 and y_step == 0:
                continue
            # 计算下一个点的坐标,即 x 值和 y 值
            next_x = self.x_values[-1] + x_step
            next_y = self.y_values[-1] + y_step
            self.x_values.append(next_x)
            self.y_values.append(next_y)
```

在上述代码中，类 RandomA 包含两个函数：__init__( )和 shibai( )，其中后者用于计算随机漫步经过的所有点。

(1) 函数 __init__( )：实现初始化处理。

▶ 为了能够做出随机决策，首先将所有可能的选择都存储在一个列表中。在每次做出具体决策时，通过"from random import choice"代码使用函数 choice( )来决定使用哪种选择。

▶ 接下来将随机漫步包含的默认点数设置为 5100，这个数值能够确保生成有趣的模式，同时也能够确保快速地模拟随机漫步。

▶ 然后在第 8 行和第 9 行代码中创建了两个用于存储 x 和 y 值的列表，并设置每次

漫步都从点(0，0)开始出发。

(2) 函数 shibai( )：功能是生成漫步包含的点，并决定每次漫步的方向。

- 第 13 行：使用 while 语句建立了一个循环，这个循环可以不断运行，直到漫步包含所需数量的点为止。这个函数的主要功能是告知 Python 应该如何模拟 4 种漫步决定：向右走还是向左走？沿指定的方向走多远？向上走还是向下走？沿选定的方向走多远？

- 第 15 行：使用 choice([1, −1])给 x_direction 设置一个值，在漫步时要么表示向右走的 1，要么表示向左走的−1。

- 第 16 行：使用 choice([0, 1, 2, 3, 4])随机地选择一个 0~4 之间的整数，告诉 Python 沿指定的方向走的距离(x_distance)。通过包含 0，不但可以沿两个轴进行移动，而且还可以沿着 y 轴进行移动。

- 第 17 行到第 20 行，将移动方向乘以移动距离，以确定沿 x 轴移动的距离。如果 x_step 为正则向右移动，如果为负则向左移动，如果为 0 则垂直移动；如果 y_step 为正则向上移动，如果为负则向下移动，如果为零则水平移动。

- 第 22 行和第 23 行：开始执行下一次循环。如果 x_step 和 y_step 都为零则原地踏步，在程序中必须杜绝这种原地踏步的情况发生。

- 第 25~28 行：为了获取漫步中下一个点的 x 值，将 x_step 与 x_values 中的最后一个值相加，对 y 值进行相同的处理。获得下一个点的 x 值和 y 值之后，将它们分别附加到列表 x_values 和 y_values 的末尾。

2. 在 Python 程序中绘制随机漫步图

在前面的实例文件 random_walk.py 中已经创建名为 RandomA 的类，在下面的实例文件 yun.py 中，将借助 matplotlib 将类 RandomA 中生成的漫步数据绘制出来，最终生成一个随机漫步图。

源码路径：codes\3\3-3\yun.py

```
import matplotlib.pyplot as plt
from random_walk import RandomA
#只要当前程序是活动的,就要不断模拟随机漫步过程
while True:
    #创建一个随机漫步实例,将包含的点都绘制出来
    rw = RandomA(51000)
    rw.shibai()
    # 设置绘图窗口的尺寸大小
    plt.figure(dpi=128, figsize=(10, 6))
    point_numbers = list(range(rw.num_points))
    plt.scatter(rw.x_values, rw.y_values, c=point_numbers, cmap=plt.cm.Blues,
        edgecolors='none', s=1)
    # 用特别的样式(红色、绿色和粗点)突出起点和终点
    plt.scatter(0, 0, c='green', edgecolors='none', s=100)
    plt.scatter(rw.x_values[-1], rw.y_values[-1], c='red', edgecolors='none',
        s=100)
    # 隐藏坐标轴
```

```
plt.axes().get_xaxis().set_visible(False)
plt.axes().get_yaxis().set_visible(False)
plt.show()
keep_running = input("哥，还继续漫步吗？ (y/n): ")
if keep_running == 'n':
    break
```

(1) 第 1 行和第 2 行：分别导入模块 pyplot 和前面编写的类 RandomA。

(2) 第 6 行：创建一个 RandomA 实例，将其存储到 rw 中，并设置点的数目。

(3) 第 7 行：调用函数 shibai( )。

(4) 第 9 行：使用函数 figure( )设置图表的宽度、高度和分辨率。

(5) 第 10 行：使用颜色映射来指出漫步中各点的先后顺序，并删除每个点的黑色轮廓，这样可以让它们的颜色显示更加明显。为了根据漫步中各点的先后顺序进行着色，需要传递参数 c，并将其设置为一个列表，其中包含各点的先后顺序。由于这些点是按顺序绘制的，因此给参数 c 指定的列表只需包含数字 1~51000 即可。使用函数 range( )生成一个数字列表，其中包含的数字个数与漫步包含的点数相同。接下来，将这个列表存储在 point_numbers 中，以便后面使用它来设置每个漫步点的颜色。

(6) 第 11、12 行：将随机漫步包含的 x 和 y 值传递给函数 scatter( )，并选择了合适的点尺寸。将参数 c 设置为在第 10 行中创建的 point_numbers，用于设置使用颜色映射 Blues，并传递实参 edgecolors='none'删除每个点周围的轮廓。

(7) 第 14~16 行：在绘制随机漫步图后重新绘制起点和终点，目的是突出显示随机漫步过程中的起点和终点。在程序中让起点和终点变得更大，并用不同的颜色显示。为了实现突出显示的功能，使用绿色绘制点(0,0)，设置这个点比其他的点都粗大(设置为 s=100)。在突出显示终点时，在漫步包含的最后一个坐标的 x 值和 y 值的位置绘制一个点，并设置它的颜色是红色，将其粗大值 s 设置为 100。

(8) 第 18、19 行：隐藏图表中的坐标轴，使用函数 plt.axes( )将每条坐标轴的可见性都设置为 False。

(9) 第 21~23 行：实现模拟多次随机漫步功能，因为每次随机漫步都不同，要想在不多次运行程序的情况下使用前面的代码实现模拟多次随机漫步的功能，最简单的办法是将这些代码放在一个 while 循环中。这样通过本实例模拟一次随机漫步后，在 matplotlib 查看器中可以浏览漫步结果，接下来可以在不关闭查看器的情况下暂停程序的执行，并询问是否要再模拟一次随机漫步。如果输入 y 则可以模拟多次随机漫步。这些随机漫步都在起点附近进行，大多数是沿着特定方向偏离起点，漫步点分布不均匀等。要结束程序的运行，只需输入 n 即可。

本实例最终执行后的效果如图 3-26 所示。

图 3-26　执行后的效果

3.3.2　大数据可视化分析某地的天气情况

假设存在一个 CSV 文件 death_valley_2014.csv，里面保存了 2014 年某地全年每一天各个时段的温度，下面开始可视化分析这个 CSV 文件。

(1) 可视化展示 2014 年 4 月的温度。

编写 Python 文件 4month.py，使用库 Matplotlib 绘制统计折线图，可视化显示 2014 年 4 月的温度。文件 4month.py 的具体实现代码如下：

```
import csv
from datetime import datetime
from matplotlib import pyplot as plt
plt.rcParams['font.sans-serif'] = ['SimHei'] # 指定默认字体
plt.rcParams['axes.unicode_minus'] = False # 解决保存图像是负号'-'显示为方块的问题
filename='./csv/death_valley_2014.csv'
with open(filename,'r')as file:
    #1.创建阅读器对象
    reader=csv.reader(file)
    #2.读取文件头信息
    header_row=next(reader)

    #3.保存最高气温数据
    dates,hights=[],[]
    for row in reader:
        current_date=datetime.strptime(row[0],"%Y-%m-%d")
        dates.append(current_date)
        #4.将字符串转换为整型数据
        hights.append(row[1])
    #5.根据数据绘制图形
    fig=plt.figure(dpi=128,figsize=(10,6))
    #6.将列表 hights 传给 plot()方法
    plt.plot(dates,hights,c='red')
    #7.设置图形的格式
```

```
plt.title('2014年4月份的温度',fontsize=24)
plt.xlabel('',fontsize=26)
# 8.绘制斜线日期标签
fig.autofmt_xdate()
plt.ylabel('华摄氏度F',fontsize=16)
plt.tick_params(axis='both',which='major',labelsize=16)
plt.show()
```

本实例的执行效果如图 3-27 所示。

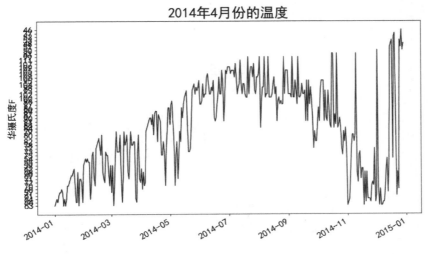

图 3-27　执行效果

(2) 可视化展示全年天气数据。

编写 Python 文件 year.py，使用库 Matplotlib 绘制统计折线图，可视化显示 2014 年全年的温度。文件 year.py 的具体实现代码如下：

```
import csv
from datetime import datetime
from matplotlib import pyplot as plt
plt.rcParams['font.sans-serif'] = ['SimHei'] # 指定默认字体
plt.rcParams['axes.unicode_minus'] = False # 解决保存图像是负号'-'显示为方块的问题
filename='./csv/death_valley_2014.csv'
with open(filename,'r')as file:
    #1.创建阅读器对象
    reader=csv.reader(file)
    #2.读取文件头信息
    header_row=next(reader)
    
    #3.保存最高气温数据
    dates,hights=[],[]
    for row in reader:
        current_date=datetime.strptime(row[0],"%Y-%m-%d")
        dates.append(current_date)
        #4.将字符串转换为整型数据
        hights.append(row[1])
    #5.根据数据绘制图形
```

```
fig=plt.figure(dpi=128,figsize=(10,6))
#6.将列表hights传给plot()方法
plt.plot(dates,hights,c='red')
#7.设置图形的格式
plt.title('2014全年的温度',fontsize=24)
plt.xlabel('',fontsize=26)
# 8.绘制斜线日期标签
fig.autofmt_xdate()
plt.ylabel('华摄氏度F',fontsize=16)
plt.tick_params(axis='both',which='major',labelsize=16)
plt.show()
```

本实例的执行效果如图3-28所示。

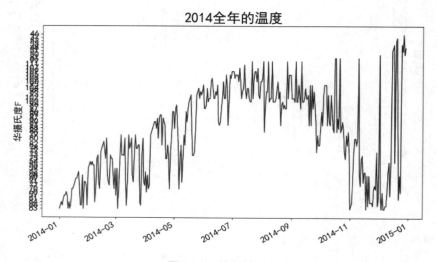

图3-28　执行效果

(3) 可视化展示某年最高温度和最低温度。

编写Python文件high_lows.py，使用库Matplotlib绘制统计折线图，统计出2014年的最高温度和最低温度。文件high_lows.py的具体实现代码如下。

源码路径：codes\3\3-3\high_lows.py

```
import csv
from matplotlib import pyplot as plt
from datetime import datetime

file = './csv/death_valley_2014.csv'
with open(file) as f:
    reader = csv.reader(f)
    header_row = next(reader)
    # 从文件中获取最高气温
    highs,dates,lows = [], [], []
    for row in reader:
        try:
            date = datetime.strptime(row[0],"%Y-%m-%d")
            high = int(row[1])
```

```
            low = int(row[3])
        except ValueError:
            print(date,'missing data')
        else:
            highs.append(high)
            dates.append(date)
            lows.append(low)

# 根据数据绘制图形
fig = plt.figure(figsize=(10,6))
plt.plot(dates,highs,c='r',alpha=0.5)
plt.plot(dates,lows,c='b',alpha=0.5)
plt.fill_between(dates,highs,lows,facecolor='b',alpha=0.2)
## 设置图形的格式
plt.title('Daily high and low temperatures-2014',fontsize=16)
plt.xlabel('',fontsize=12)
fig.autofmt_xdate()
plt.ylabel('Temperature(F)',fontsize=12)
plt.tick_params(axis='both',which='major',labelsize=20)
plt.show()
```

执行后的效果如图 3-29 所示。

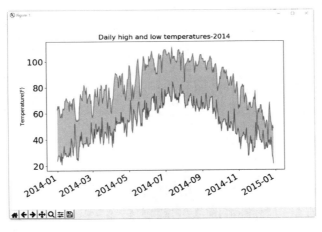

图 3-29　执行后的效果

3.3.3　在 Tkinter 中使用 Matplotlib 绘制图表

在下面的实例文件 123.py 中，演示了在标准 GUI 程序 Tkinter 中使用 Matplotlib 绘制图表的过程。

源码路径：codes\3\3-3\123.py

```
class App(tk.Tk):
    def __init__(self, parent=None):
        tk.Tk.__init__(self, parent)
        self.parent = parent
        self.initialize()
```

```python
    def initialize(self):
        self.title("在Tkinter中使用Matplotlib! ")
        button = tk.Button(self, text="退出", command=self.on_click)
        button.grid(row=1, column=0)
        self.mu = tk.DoubleVar()
        self.mu.set(5.0)   #参数的默认值是"mu"
        slider_mu = tk.Scale(self,
                    from_=7, to=0, resolution=0.1,
                    label='mu', variable=self.mu,
                    command=self.on_change
                    )
        slider_mu.grid(row=0, column=0)
        self.n = tk.IntVar()
        self.n.set(512)   #参数的默认值是"n"
        slider_n = tk.Scale(self,
                    from_=512, to=2,
                    label='n', variable=self.n, command=self.on_change
                    )
        slider_n.grid(row=0, column=1)

        fig = Figure(figsize=(6, 4), dpi=96)
        ax = fig.add_subplot(111)
        x, y = self.data(self.n.get(), self.mu.get())
        self.line1, = ax.plot(x, y)
        self.graph = FigureCanvasTkAgg(fig, master=self)
        canvas = self.graph.get_tk_widget()
        canvas.grid(row=0, column=2)

    def on_click(self):
        self.quit()

    def on_change(self, value):
        x, y = self.data(self.n.get(), self.mu.get())
        self.line1.set_data(x, y)   # 更新data数据
        # 更新graph
        self.graph.draw()

    def data(self, n, mu):
        lst_y = []
        for i in range(n):
            lst_y.append(mu * random.random())
        return range(n), lst_y

if __name__ == "__main__":
    app = App()
    app.mainloop()
```

执行后可以拖动左侧的滑动条改变参数查看绘制的图表，执行效果如图3-30所示。

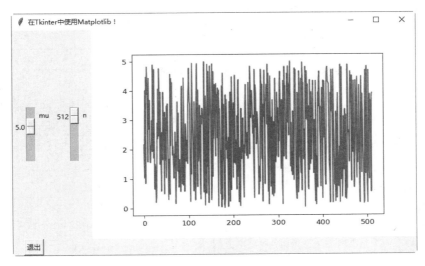

图 3-30　执行效果

3.3.4　绘制包含点、曲线、注释和箭头的统计图

请看下面的实例文件 jian.py，功能是使用 Matplotlib 在绘制的曲线中添加注释和箭头。

 源码路径：　codes\3\3-3\jian.py

```
import matplotlib.pyplot as plt    # 导入绘图模块
import numpy as np   # 导入需要生成数据的 numpy 模块

'''
添加注释   annotate()
    参数：(1)x ：注释文本
         (2)xy:
         (3) xytext:
         (4) 设置箭头,arrowprops
              arrowprops ：是一个 dict (字典)
         第一种方式:{'width':宽度,'headwidth':箭头宽,'headlength':箭头长,
                   'shrink':两端收缩总长度分数}
              例如：
arrowprops={'width':5,'headwidth':10,'headlength':10,'shrink':0.1}
         第二种方式: 'arrowstyle':样式
              例如：
              有关arrowstyle的样式: '-'、'->'、'<-'、'-['、'|-|'、'-|>'、'<|-'、'<->'
                                  'fancy','simple','wedge'
'''
x = np.random.randint(0,30,size=10)
x[5] = 30   # 把索引为 5 的位置改为 30
plt.figure(figsize=(12,6))
plt.plot(x)
plt.ylim([-2,35]) # 设置 y 轴的刻度
plt.annotate(s='this point is important',xy=(5,30),xytext=(7,31),
        arrowprops={'arrowstyle':'->'})
plt.show()
```

执行效果如图 3-31 所示。

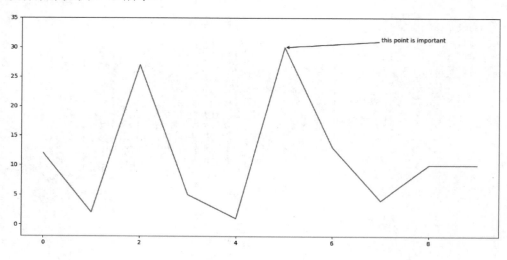

图 3-31 执行效果

再看下面的实例文件 zong.py，功能是使用 Matplotlib 绘制包含点、曲线、注释和箭头的统计图。

源码路径：codes\3\3-3\zong.py

```
from matplotlib import pyplot as plt
import numpy as np

# 绘制曲线
x = np.linspace(2, 21, 20)  # 取闭区间[2, 21]之间的等差数列,列表长度20
y = np.log10(x) + 0.5
plt.figure()  # 添加一个窗口.如果只显示一个窗口,可以省略该句
plt.plot(x, y)  # plot 在一个 figure 窗口中添加一个图,绘制曲线,默认颜色

# 绘制离散点
plt.plot(x, y, '.y')  # 绘制黄色的点,为了和曲线颜色不一样
x0, y0 = 15, np.log10(15) + 0.5
plt.annotate('Interpolation point', xy=(x0, y0), xytext=(x0, y0 - 1),
arrowprops=dict(arrowstyle='->'))  # 添加注释
for x0, y0 in zip(x, y):
    plt.quiver(x0, y0 - 0.3, 0, 1, color='g', width=0.005)  # 绘制箭头

x = range(2, 21, 5)
y = np.log10(x) + 0.5
plt.plot(x, y, 'om')  # 绘制紫红色的圆形的点
x0, y0 = 7, np.log10(7) + 0.5
plt.annotate('Original point', xy=(x0, y0), xytext=(x0, y0 - 1),
arrowprops=dict(arrowstyle='->'))
for x0, y0 in zip(x, y):
    plt.quiver(x0, y0 + 0.3, 0, -1, color='g', width=0.005)  # 绘制箭头

# 设置坐标范围
plt.xlim(2, 21)  # 设置 x 轴范围
```

```
plt.xticks(range(0, 23, 2))   # 设置x轴坐标点的值,为[0, 22]之间的以2为差值的等差数组
plt.ylim(0, 3)   # 设置y轴范围

# 显示图形
plt.show()   # 显示绘制出的图
```

对上述代码的具体说明如下。

(1) 导入 matplotlib 模块的 pyplot 类,这里主要用了 pyplot 里的一些方法。导入 numpy 用于生成一些数列,分别给 pyplot 和 numpy 记个简洁的别名 plt 和 np。

(2) 通过方法 np.linspace(start, stop, num)生成闭区间[stop, stop]里的数组长度为 num 的等差数列,在本例中想作为插值点显示出来。

(3) 通过方法 plt.figure( )添加窗口。如果把所有图形绘制在一个窗口里,则可以省略该行代码,因为 figure(1)会被默认创建。如果想添加窗口,则需要再添加一行代码 plt.figure( ),plt.figure(num)的窗口序号 num 会自动增加 1。

(4) 通过方法 plt.plot( )在窗口中绘制曲线,传递 x、y 参数,分别表示横轴和纵轴。

本实例的执行效果如图 3-32 所示。

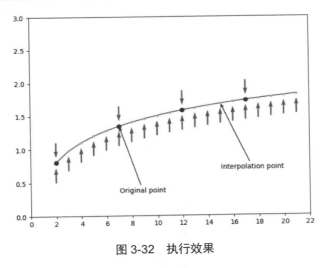

图 3-32 执行效果

3.3.5 在两栋房子之间绘制箭头指示符

请看下面的实例文件 zhishi.py,首先使用散点图表示房子的位置,然后在两栋房子之间绘制箭头指示符。

源码路径:codes\3\3-3\zhishi.py

```
import numpy as np
import matplotlib.pyplot as plt

fig, ax = plt.subplots(figsize=(5, 5))
ax.set_aspect(1)

x1 = -1 + np.random.randn(100)
```

```python
y1 = -1 + np.random.randn(100)
x2 = 1. + np.random.randn(100)
y2 = 1. + np.random.randn(100)

ax.scatter(x1, y1, color="r")
ax.scatter(x2, y2, color="g")

bbox_props = dict(boxstyle="round", fc="w", ec="0.5", alpha=0.9)
ax.text(-2, -2, "住宅 A", ha="center", va="center", size=20,
        bbox=bbox_props, fontproperties="SimSun")
ax.text(2, 2, "住宅 B", ha="center", va="center", size=20,
        bbox=bbox_props,fontproperties="SimSun")

bbox_props = dict(boxstyle="rarrow", fc=(0.8, 0.9, 0.9), ec="b", lw=2)
t = ax.text(0, 0, "Direction", ha="center", va="center", rotation=45,
        size=15,
        bbox=bbox_props)

bb = t.get_bbox_patch()
bb.set_boxstyle("rarrow", pad=0.6)

ax.set_xlim(-4, 4)
ax.set_ylim(-4, 4)

plt.show()
```

本实例的执行效果如图 3-33 所示。

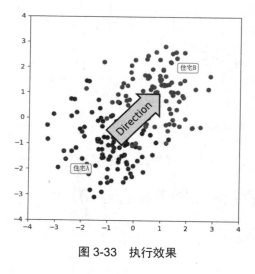

图 3-33　执行效果

3.3.6　根据坐标绘制行走路线图

请看下面的实例文件 road.py，功能是根据预先设置的位置坐标绘制行走路线图。

源码路径：codes\3\3-3\road.py

```
import matplotlib.pyplot as plt
import numpy
```

```python
import matplotlib.colors as colors
import matplotlib.cm as cmx

_locations = [
    (4, 4),  # depot
    (4, 4),  # unload depot_prime
    (4, 4),  # unload depot_second
    (4, 4),  # unload depot_fourth
    (4, 4),  # unload depot_fourth
    (4, 4),  # unload depot_fifth
    (2, 0),
    (8, 0),  # locations to visit
    (0, 1),
    (1, 1),
    (5, 2),
    (7, 2),
    (3, 3),
    (6, 3),
    (5, 5),
    (8, 5),
    (1, 6),
    (2, 6),
    (3, 7),
    (6, 7),
    (0, 8),
    (7, 8)
]

plt.figure(figsize=(10, 10))
p1 = [l[0] for l in _locations]
p2 = [l[1] for l in _locations]
plt.plot(p1[:6], p2[:6], 'g*', ms=20, label='depot')
plt.plot(p1[6:], p2[6:], 'ro', ms=15, label='customer')
plt.grid(True)
plt.legend(loc='lower left')

way = [[0, 12, 18, 17, 16, 4, 14, 10, 11, 13, 5], [0, 6, 9, 8, 20, 3], [0, 19, 21, 15, 7, 2]]   #

cmap = plt.cm.jet
cNorm = colors.Normalize(vmin=0, vmax=len(way))
scalarMap = cmx.ScalarMappable(norm=cNorm,cmap=cmap)

for k in range(0, len(way)):
    way0 = way[k]
    colorVal = scalarMap.to_rgba(k)
    for i in range(0, len(way0)-1):
        start = _locations[way0[i]]
        end = _locations[way0[i+1]]
#        plt.arrow(start[0], start[1], end[0]-start[0], end[1]-start[1],
length_includes_head=True,
#                  head_width=0.2, head_length=0.3, fc='k', ec='k', lw=2,
ls=lineStyle[k], color='red')
        plt.arrow(start[0], start[1], end[0]-start[0], end[1]-start[1],
                  length_includes_head=True, head_width=0.2, lw=2,
```

```
        color=colorVal)
plt.show()
```

在上述代码中，使用_locations 保存了表示位置的坐标，使用 cmap 表示绘制的颜色库，通过 cNorm 设置颜色的范围，有几条线路就设置几种颜色，scalarMap 表示颜色生成完毕。在绘图时根据索引获得相应的颜色。本实例的执行效果如图 3-34 所示。

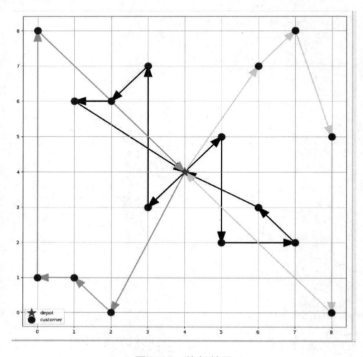

图 3-34　执行效果

3.3.7　绘制方程式曲线图

请看下面的实例文件 xian.py，功能是使用库 Matplotlib 绘制以下两个方程式的曲线图。

$$f(t)=e^{-t}\cos(2\pi t)$$
$$g(t)=\sin(2\pi t)\cos(3\pi t)$$

源码路径：codes\3\3-3\xian.py

```
import numpy as np
import matplotlib.pyplot as plt

def f1(t):
    return np.exp(-t) * np.cos(2 * np.pi * t)

def f2(t):
    return np.sin(2 * np.pi * t) * np.cos(3 * np.pi * t)

t = np.arange(0.0, 5.0, 0.02)
```

```python
plt.figure(figsize=(8, 7), dpi=98)
p1 = plt.subplot(211)
p2 = plt.subplot(212)

label_f1 = "$f(t)=e^{-t} \cos (2 \pi t)$"
label_f2 = "$g(t)=\sin (2 \pi t) \cos (3 \pi t)$"

p1.plot(t, f1(t), "g-", label=label_f1)
p2.plot(t, f2(t), "r-.", label=label_f2, linewidth=2)

p1.axis([0.0, 5.01, -1.0, 1.5])

p1.set_ylabel("v", fontsize=14)
p1.set_title("A simple example", fontsize=18)
p1.grid(True)
# p1.legend()

tx = 2
ty = 0.9
p1.text(tx, ty, label_f1, fontsize=15, verticalalignment="top",
horizontalalignment="right")

p2.axis([0.0, 5.01, -1.0, 1.5])
p2.set_ylabel("v", fontsize=14)
p2.set_xlabel("t", fontsize=14)
# p2.legend()
tx = 2
ty = 0.9
p2.text(tx, ty, label_f2, fontsize=15, verticalalignment="bottom",
horizontalalignment="left")

p2.annotate('', xy=(1.8, 0.5), xytext=(tx, ty),
arrowprops=dict(arrowstyle="->", connectionstyle="arc3"))

plt.show()
```

本实例的执行效果如图 3-35 所示。

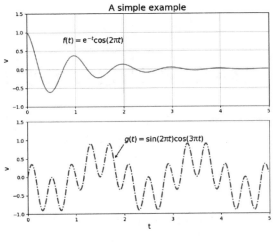

图 3-35　执行效果

3.3.8 绘制星空图

请看下面的实例文件 cloud.py，功能是使用库 Matplotlib 绘制不同样式的小星星。

源码路径：codes\3\3-3\cloud.py

```
import matplotlib.pyplot as plt    # 导入绘图模块
import numpy as np   # 导入需要生成数据的numpy 模块

x = np.random.randn(100)
y = np.random.randn(100)
'''设置每个点的颜色随机生成'''
color = np.random.random(300).reshape((100, 3))   # 一千行三列
'''设置每个点的大小随机生成'''
size = np.random.randint(0, 100, 100)
plt.scatter(x, y, color=color, s=size, marker='*')
plt.show()
```

执行后的效果如图 3-36 所示。

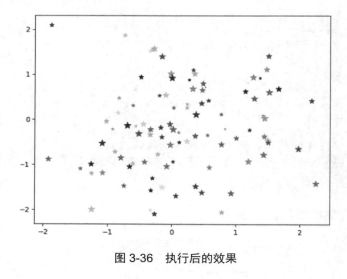

图 3-36　执行后的效果

3.4　绘制 BTC(比特币)和 ETH(以太币)的价格走势图

在本节的内容中，将远程获取当前国际市场中 BTC(比特币)和 ETH(以太币)的实时价格，并绘制 BTC 和 ETH 的价格走势曲线图。

3.4.1　抓取数据

编写实例文件 Assignment_Step1.py，功能是抓取权威网站中 BTC 和 ETH 的报价数据，并打印输出显示 BTC 和 ETH 的当前价格。

扫码观看本节视频讲解

源码路径：codes\3\3-4\Crypto-Analysis\Assignment_Step1.py

```python
import requests
def price(symbol, comparison_symbols=['USD'], exchange=''):
    url = 'https://min-api.cryptocompare.com/data/price?fsym={}&tsyms={}'\
            .format(symbol.upper(), ','.join(comparison_symbols).upper())
    if exchange:
        url += '&e={}'.format(exchange)
    page = requests.get(url)
    data = page.json()
    return data

print("当前 BTC 的美元价格为："+str(price('BTC')))
print("当前 ETH 的美元价格为："+str(price('ETH')))
```

2020 年 5 月 13 日执行本实例程序后会输出：

```
当前 BTC 的美元价格为：{'USD': 8910.06}
当前 ETH 的美元价格为：{'USD': 190.47}
```

3.4.2 绘制 BTC/美元价格曲线

编写实例文件 Assignment_Step2.py，功能是根据当前的 BTC 价格，使用库 Matplotlib 绘制 BTC/美元价格曲线图。

源码路径：codes\3\3-4\Crypto-Analysis\Assignment_Step2.py

```python
from Assignment_Step1 import price
import datetime
import matplotlib.pyplot as plt

x=[0]
y=[0]
fig = plt.gcf()
fig.show()
fig.canvas.draw()
plt.ylim([0, 20000])
i=0
while(True):
    data = price('BTC')
    i+=1
    x.append(i)
    y.append(data['USD'])
    plt.title("BTC vs USD, Last Update is: "+str(datetime.datetime.now()))
    plt.plot(x,y)
    fig.canvas.draw()
    plt.pause(1000)
```

执行后的效果如图 3-37 所示。

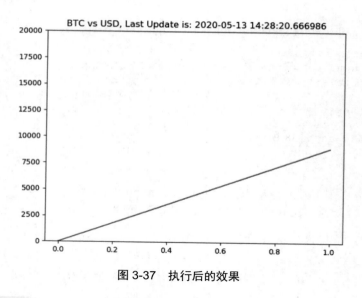

图 3-37 执行后的效果

3.4.3 绘制 BTC 和 ETH 的历史价格曲线图

编写实例文件 Assignment_Step3.py，功能是首先爬取权威网站中的 BTC 和 ETH 的历史价格数据，然后使用库 Matplotlib 和 pandas 绘制 BTC 和 ETH 的历史价格曲线图。

源码路径：codes\3\3-4\Crypto-Analysis\Assignment_Step3.py

```python
import requests
import datetime
import pandas as pd
import matplotlib.pyplot as plt

def hourly_price_historical(symbol, comparison_symbol, limit, aggregate, exchange=''):
    url = 'https://min-api.cryptocompare.com/data/histohour?fsym={}&tsym={}&limit={}&aggregate={}'\
        .format(symbol.upper(), comparison_symbol.upper(), limit, aggregate)
    if exchange:
        url += '&e={}'.format(exchange)
    print(url)
    page = requests.get(url)
    data = page.json()['Data']
    df = pd.DataFrame(data)
    df['timestamp'] = [datetime.datetime.fromtimestamp(d) for d in df.time]

    return df

def plotchart(axis, df, symbol, comparison_symbol):
    axis.plot(df.timestamp, df.close)

df1 = hourly_price_historical('BTC','USD', 2000, 1)
df2 = hourly_price_historical('ETH','USD', 2000, 1)
```

```
f, axarr = plt.subplots(2)

plotchart(axarr[0],df1,'BTC','USD')
plotchart(axarr[1],df2,'ETH','USD')

plt.show()
```

执行后的效果如图 3-38 所示。

图 3-38　执行后的效果

3.5　Flask+pygal+SQLite 实现数据分析

在本节的内容中，将使用 Flask+pygal+SQLite3 实现数据分析功能。将需要分析的数据保存在 SQLite 3 数据库中，然后在 Flask Web 网页中使用库 pygal 绘制出对应的统计图。

3.5.1　创建数据库

扫码观看本节视频讲解

首先使用 PyCharm 创建 Flask Web 项目，然后通过文件 models.py 设计 SQLite 数据库的结构，主要实现代码如下：

```
from dbconnect import db

# 许可证申请数量
class Appinfo(db.Model):
    __tablename__ = 'appinfo'
    # 注意这句,网上有些实例上并没有
    # 必须设置主键
    id = db.Column(db.Integer, primary_key=True)
```

```python
    year = db.Column(db.String(20))
    month = db.Column(db.String(20))
    cnt = db.Column(db.String(20))

    def __init__(self, year, month, cnt):
        self.year = year
        self.month = month
        self.cnt = cnt

    def __str__(self):
        return self.year + ":" + self.month + ":" + self.cnt

    def __repr__(self):
        return self.year + ":" + self.month + ":" + self.cnt

    def save(self):
        db.session.add(self)
        db.session.commit()
```

在数据库表 appinfo 中添加数据，如图 3-39 所示。

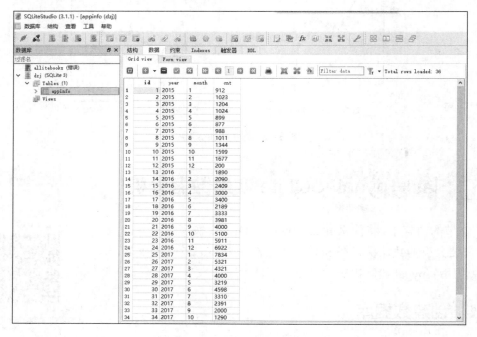

图 3-39　数据库 dzj.db 中的数据

3.5.2　绘制统计图

编写 Flask Web 启动文件 pygal_test.py，首先建立 URL 路径导航指向模板文件 index.htm，然后提取数据库中的数据，并使用 pygal 库绘制出统计图表。文件 pygal_test.py 的主要实现代码如下：

```
app = Flask(__name__)
dbpath = app.root_path
```

```python
# 注意斜线的方向
app.config['SQLALCHEMY_DATABASE_URI'] = r'sqlite:///' + dbpath + '/dzj.db'
app.config['SQLALCHEMY_TRACK_MODIFICATIONS'] = True
print(app.config['SQLALCHEMY_DATABASE_URI'])

db.init_app(app)

@app.route('/')
def APPLYTBLINFO():
    db.create_all()
    # 在第一次调用时执行就可以
    appinfos = Appinfo.query.all()
    # 选择年份
    list_year = []
    # 选择月份
    list_month = []
    # 月份对应的数字
    map_cnt = {}
    for info in appinfos:
        if info.year not in list_year:
            list_year.append(info.year)
            map_cnt[info.year] = [int(info.cnt)]
        else:
            map_cnt[info.year].append(int(info.cnt))
        if info.month not in list_month:
            list_month.append(info.month)
    line_chart = pygal.Line()
    line_chart.title = '信息'
    line_chart.x_labels = map(str, list_month)
    for year in list_year:
        line_chart.add(str(year) + "年", map_cnt[year])
    return render_template('index.html', chart=line_chart)


if __name__ == '__main__':
    app.run(debug=True)
```

模板文件 index.htm 的具体实现代码如下：

```
<body style="width: 1000px;margin: auto">
<div id="container">
    <div id="header" style="background: burlywood;height: 50px;">
        <h2 style="font-size: 30px; position: absolute; margin-top: 10px;margin-left: 300px;
        text-align:center;">数据走势图分析</h2>
    </div>
    <div id="leftbar" style="width: 200px;height: 600px;background: cadetblue;float: left">
        <h2 style="margin-left: 20px">数据图总览</h2><br/>
        <table>
            <tr>
                <td>
                    <a name="appinfo" style="margin-left: 20px;">数量分析图</a><br>
                </td>
            </tr>
```

```
        </table>
    </div>
    <div id="chart" style="width: 800px;float: left">
        <embed type="image/svg+xml" src= {{ chart.render_data_uri()|safe }} />
    </div>
</div>
</body>
```

执行 Flask Web 项目，在浏览器中输入 http://127.0.0.1:5000/后会显示绘制的统计图。执行效果如图 3-40 所示。

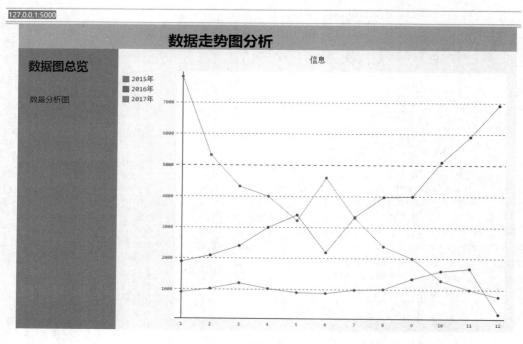

图 3-40　绘制的统计图效果

第 4 章

绘制柱状图

　　柱状图又称为柱形图、长条图、条状图或棒形图,是一种以长方形的长度为变量的统计图表。在现实应用中,经常使用柱状图来展示数据分析的结果,这样可以更加直观、易懂。本章将详细讲解使用 Python 绘制柱状图的知识。

4.1 绘制基本的柱状图

在数据可视化分析工作中，经常需要进行数据统计并绘制相关的柱状图。本节将详细讲解在 Python 程序中使用各种库绘制基本柱状图的知识。

扫码观看本节视频讲解

4.1.1 绘制只有一个柱子的柱状图

在 Python 程序中，可以使用 Matplotlib 很容易地绘制一个柱子的柱状图。例如，在线的实例文件 bar01.py 中，只需使用 3 行代码就可以绘制一个简单的柱状图。

源码路径：codes\4\4-1\bar01.py

```
import matplotlib.pyplot as plt
plt.bar(x = 0,height = 1)
plt.show()
```

在上述代码中，首先使用 import 导入 matplotlib.pyplot，然后直接调用其 bar( ) 函数绘制柱状图，最后使用 show( ) 函数显示图像。其中在函数 bar( ) 中存在以下两个参数。
- left：柱形的左边缘的位置，如果指定为 1，那么当前柱形的左边缘的 x 值就是 1.0。
- height：这是柱形的高度，也就是 y 轴的值。

执行上述代码后会绘制一个柱状图，如图 4-1 所示。

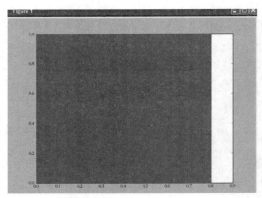

图 4-1 执行效果

4.1.2 绘制有两个柱子的柱状图

虽然通过上述代码绘制了一个柱状图，但是现实效果不够直观。在绘制函数 bar( ) 中，参数 x 和 height 除了可以使用单独的值(此时是一个柱形)外，还可以使用元组来替换(此时代表多个矩形)。例如，在下面的实例文件 bar02.py 中，演示了使用 matplotlib 绘制有两个柱子的柱状图的过程。

源码路径：codes\4\4-1\bar02.py

```
import matplotlib.pyplot as plt          #导入模块
plt.bar(x = (0,1),height = (1,0.5))      #绘制两个柱子的柱形图
plt.show()                               #显示绘制的图
```

执行后的效果如图 4-2 所示。

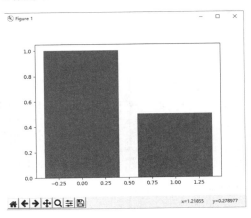

图 4-2　执行后的效果

在上述实例代码中，x = (0,1)的意思是总共有两个矩形，其中第一个矩形的左边缘为 0，第二个矩形的左边缘为 1。参数 height 的含义也是同理。当然，此时有的读者可能觉得这两个矩形"太宽"了，不够美观。此时可以通过指定函数 bar( )中的 width 参数来设置它们的宽度。请看下面的实例文件 bar03.py，功能是设置柱子的宽度 width 为 0.35。

 源码路径：codes\4\4-1\bar03.py

```
import matplotlib.pyplot as plt
plt.bar(x = (0,1),height = (1,0.5),width = 0.35)
plt.show()
```

执行后的效果如图 4-3 所示。

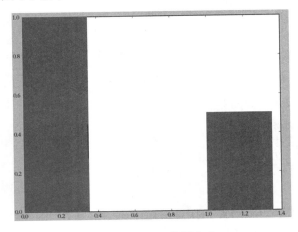

图 4-3　设置柱状图宽度

4.1.3 设置柱状图的标签

这时可能有的读者会问：需要标明 x 轴和 y 轴的说明信息，如使用 x 轴表示性别、使用 y 轴表示人数。例如，在下面的实例文件 bar04.py 中，演示了使用库 matplotlib 绘制有说明信息柱状图的过程。

源码路径：**codes\4\4-1\bar04.py**

```
import matplotlib.pyplot as plt
from pylab import *
mpl.rcParams['font.sans-serif'] = ['SimHei']           #指定默认字体
mpl.rcParams['axes.unicode_minus'] = False   #解决保存图像时负号'-'显示为方块的问题
plt.xlabel(u'性别')                                    #x 轴的说明信息
plt.ylabel(u'人数')                                    #y 轴的说明信息
plt.bar(x = (0,1),height = (1,0.5),width = 0.35)
plt.show()
```

上述代码执行后的效果如图 4-4 所示。

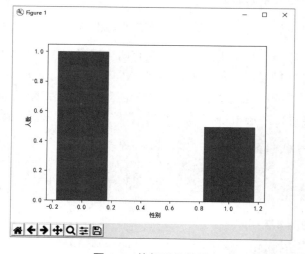

图 4-4 执行后的效果

接下来可以对 x 轴上的每个 bar 进行说明，如设置第一个柱子是"男"、第二个柱子是"女"。此时可以通过以下实例文件 bar05.py 实现。

源码路径：**codes\4\4-1\bar05.py**

```
plt.xlabel(u'性别')
plt.ylabel(u'人数')
plt.xticks((0,1),(u'男',u'女'))
plt.bar(x = (0,1),height = (1,0.5),width = 0.35)
plt.show()
```

在上述代码中，函数 plt.xticks( )的用法和前面使用的 x 和 height 的用法差不多。如果有几个 bar，那么就对应几维的元组，其中第一个参数表示文字的位置，第二个参数表示具体的文字说明。不过这里有个问题，如有时指定的位置有些"偏移"，最理想的状态应该在每个矩形的中间。可以通过直接指定函数 bar( )里面的 align="center"就可以让文字居中了。

```
plt.xlabel(u'性别')
plt.ylabel(u'人数')
plt.xticks((0,1),(u'男',u'女'))
plt.bar(x = (0,1),height = (1,0.5),width = 0.35,align="center")
plt.show()
```

此时的执行效果如图 4-5 所示。

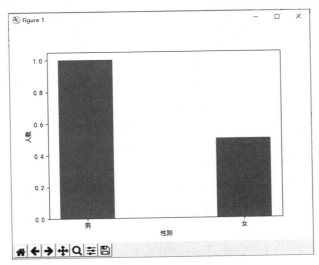

图 4-5　执行效果

接下来可以通过以下代码给柱状图表加一个标题：

```
plt.title(u"性别比例分析")
```

为了使整个程序显得更加科学、合理，接下来可以通过以下代码设置一个图例：

```
plt.xlabel(u'性别')
plt.ylabel(u'人数')
plt.title(u"性别比例分析")
plt.xticks((0,1),(u'男',u'女'))
rect = plt.bar(left = (0,1),height = (1,0.5),width = 0.35,align="center")
plt.legend((rect,),(u"图例",))
plt.show()
```

在上述代码中用到了函数 legend( )，里面的参数必须是元组。即使只有一个图例也必须是元组；否则显示不正确。此时的执行效果如图 4-6 所示。

接下来还可以在每个矩形的上面标注对应的 y 值，此时需要使用以下通用的方法实现：

```
def autolabel(rects):
    for rect in rects:
        height = rect.get_height()
        plt.text(rect.get_x()+rect.get_width()/2., 1.03*height, '%s' % float(height))
```

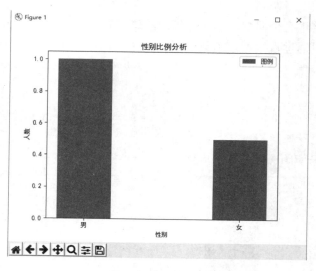

图 4-6 执行效果

在上述实例代码中，其中 plt.text 有 3 个参数，分别是 x 坐标、y 坐标、要显示的文字。调用函数 autolabel( )的具体实现代码如下：

```
autolabel(rect)
```

为了避免绘制矩形柱状图紧靠着顶部，最好能够空出一段距离，此时可以通过函数 bar( ) 的属性参数 yerr 来设置。一旦设置了这个参数，对应的矩形上面就会有一个竖着的线。当把 yerr 值设置得很小时，上面的空白就自动空出来了。

```
rect = plt.bar(left = (0,1),height = (1,0.5),width = 0.35,align="center",yerr=0.0001)
```

至此，一个比较美观的柱状图绘制完毕，将代码整理并保存在以下实例文件 bar06.py 中，具体实现代码如下。

源码路径：codes\4\4-1\bar06.py

```python
import matplotlib.pyplot as plt
from pylab import *
mpl.rcParams['font.sans-serif'] = ['SimHei']    #指定默认字体
mpl.rcParams['axes.unicode_minus'] = False   #解决保存图像时负号'-'显示为方块的问题
def autolabel(rects):
    for rect in rects:
        height = rect.get_height()
        plt.text(rect.get_x()+rect.get_width()/2., 1.03*height, '%s' % float(height))
plt.xlabel(u'性别')
plt.ylabel(u'人数')
plt.title(u"性别比例分析")
plt.xticks((0,1),(u'男',u'女'))
#绘制柱形图
rect = plt.bar(x = (0,1),height = (1,0.5),width = 0.35,align="center",yerr=0.0001)
plt.legend((rect,),(u"图例",))
autolabel(rect)
plt.show()
```

上述代码执行后的效果如图 4-7 所示。

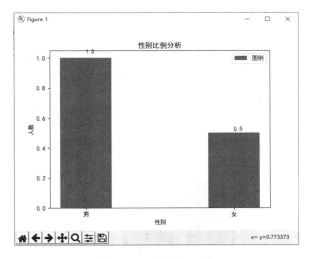

图 4-7　执行后的效果

4.1.4　设置柱状图的颜色

请看下面的实例文件 bar06-1.py，功能是使用库 matplotlib 绘制绿颜色的柱状图。

 源码路径：codes\4\4-1\bar06-1.py

```
import matplotlib.pyplot as plt

data = [5, 20, 15, 25, 10]
plt.bar(range(len(data)), data, fc='g')
plt.show()
```

执行后的效果如图 4-8 所示。

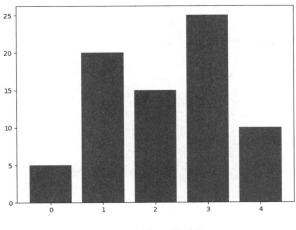

图 4-8　执行后的效果

请看下面的实例文件 bar06-2.py，功能是使用库 matplotlib 绘制不同颜色的柱状图。

源码路径：codes\4\4-1\bar06-2.py

```
import matplotlib.pyplot as plt

data = [5, 20, 15, 25, 10]
plt.bar(range(len(data)), data, color='rgb')  # or 'color=['r', 'g', 'b']'
plt.show()
```

执行后的效果如图 4-9 所示。

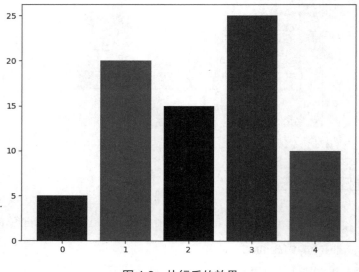

图 4-9　执行后的效果

4.1.5　绘制堆叠柱状图

请看下面的实例文件 bar06-3.py，功能是使用库 matplotlib 绘制堆叠柱状图。

源码路径：codes\4\4-1\bar06-3.py

```
import numpy as np
import matplotlib.pyplot as plt

size = 5
x = np.arange(size)
a = np.random.random(size)
b = np.random.random(size)
plt.bar(x, a, label='a')
plt.bar(x, b, bottom=a, label='b')
plt.legend()
plt.show()
```

执行后的效果如图 4-10 所示。

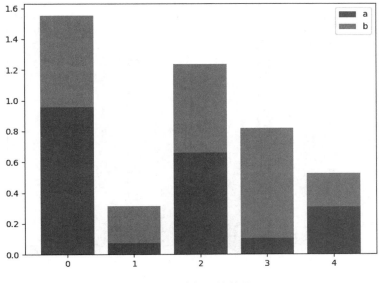

图 4-10 执行后的效果

4.1.6 绘制并列柱状图

请看下面的实例文件 bar06-4.py，功能是使用库 matplotlib 绘制并列柱状图。

源码路径：codes\4\4-1\bar06-4.py

```python
import numpy as np
import matplotlib.pyplot as plt

size = 5
x = np.arange(size)
a = np.random.random(size)
b = np.random.random(size)
c = np.random.random(size)

total_width, n = 0.8, 3
width = total_width / n
x = x - (total_width - width) / 2

plt.bar(x, a, width=width, label='a')
plt.bar(x + width, b, width=width, label='b')
plt.bar(x + 2 * width, c, width=width, label='c')
plt.legend()
plt.show()
```

执行后的效果如图 4-11 所示。

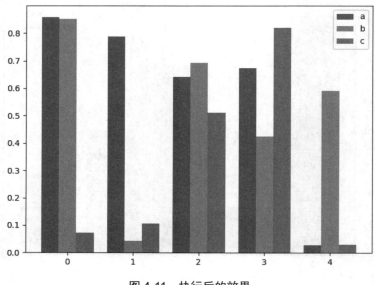

图 4-11 执行后的效果

4.1.7 绘制 2002—2013 年网页浏览器使用变化柱状图

使用库 pygal 绘制柱状图的方法十分简单，只需调用库 pygal 中的 Bar( )方法即可。例如，在下面的实例文件 bar07.py 中，绘制了 2002—2013 年网页浏览器的使用变化数据的柱状图。

源码路径：codes\4\4-1\bar07.py

```
import pygal

line_chart = pygal.Bar()
line_chart.title = '网页浏览器的使用变化(in %)'
line_chart.x_labels = map(str, range(2002, 2013))
line_chart.add('Firefox', [None, None, 0, 16.6,  25,  31, 36.4, 45.5, 46.3, 42.8, 37.1])
line_chart.add('Chrome',  [None, None, None, None, None, None,  0, 3.9, 10.8, 23.8, 35.3])
line_chart.add('IE',      [85.8, 84.6, 84.7, 74.5,  66, 58.6, 54.7, 44.8, 36.2, 26.6, 20.1])
line_chart.add('Others',  [14.2, 15.4, 15.3, 8.9,   9, 10.4, 8.9, 5.8, 6.7, 6.8, 7.5])
line_chart.render_to_file('bar_chart.svg')
```

执行后会创建生成条形图文件 bar_chart.svg，打开后的效果如图 4-12 所示。

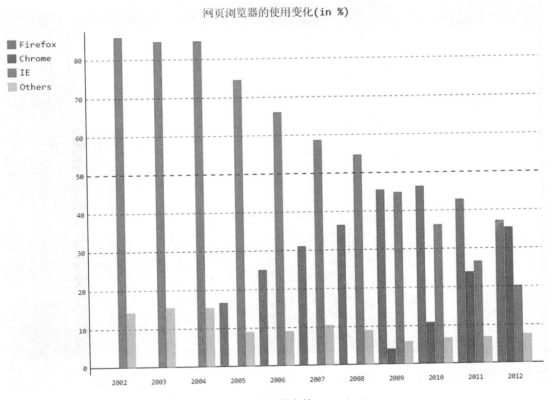

图 4-12　生成的条形图文件 bar_chart.svg

4.1.8　绘制直方图

使用库 pygal 绘制直方图的方法十分简单，只需调用库 pygal 中的 Histogram( )方法即可。例如，在下面的实例文件 bar08.py 中，使用 Histogram( )方法分别绘制宽直方图和窄直方图。

> 源码路径：codes\4\4-1\bar08.py

```
import pygal
hist = pygal.Histogram()
hist.add('Wide bars', [(5, 0, 10), (4, 5, 13), (2, 0, 15)])
hist.add('Narrow bars', [(10, 1, 2), (12, 4, 4.5), (8, 11, 13)])
hist.render_to_file('bar_chart.svg')
```

执行后会创建生成直方图文件 bar_chart.svg，打开后的效果如图 4-13 所示。

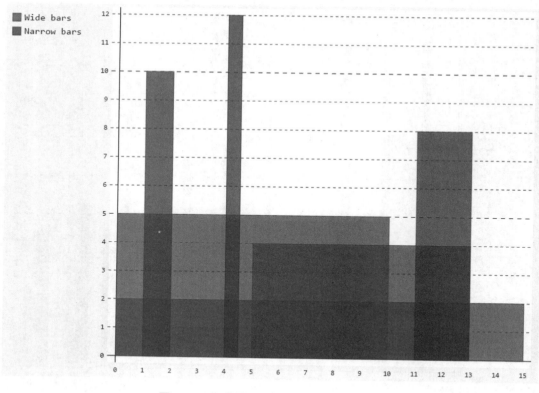

图 4-13　生成的直方图文件 bar_chart.svg

4.1.9　绘制横向柱状图

在 Python 绘图库 matplotlib 中，通过使用库函数 barh( )可以绘制横向柱形图，该函数的使用方法与常见的纵向柱形图绘制函数 bar( )的用法相似。请看下面的实例文件 bar09.py，功能是使用库 matplotlib 绘制横向柱状图。

源码路径：codes\4\4-1\bar09.py

```
import matplotlib.pyplot as plt
from matplotlib.font_manager import FontProperties

#数据
name=['1','2','3','4']
colleges=[91,34,200,100]

#图像绘制
fig,ax=plt.subplots()
b=ax.barh(range(len(name)),colleges,color='#6699CC')

#添加数据标签
for rect in b:
    w=rect.get_width()
    ax.text(w,rect.get_y()+rect.get_height()/2,'%d'%int(w),
ha='left',va='center')
```

```
#设置Y轴刻度线标签
ax.set_yticks(range(len(name)))
#font=FontProperties(fname=r'/Library/Fonts/Songti.ttc')
ax.set_yticklabels(name)

plt.show()
```

执行后的效果如图 4-14 所示。

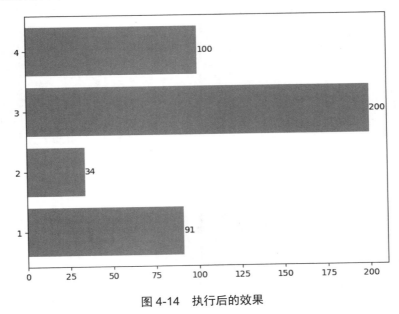

图 4-14 执行后的效果

4.1.10 绘制有图例横向柱状图

请看下面的实例文件 bar10.py，功能是使用库 matplotlib 绘制有图例横向柱状图。

源码路径：**codes\4\4-1\bar10.py**

```
import matplotlib.pyplot as plt

dic = {'a': 22, 'b': 10, 'c': 6, 'd': 4, 'e': 2, 'f': 10, 'g': 24, 'h': 16, 'i': 1, 'j': 12}
s = sorted(dic.items(), key=lambda x: x[1], reverse=False)  # 对dict 按照value 排序 True 表示翻转 ,转为了列表形式
print(s)
x_x = []
y_y = []
for i in s:
    x_x.append(i[0])
    y_y.append(i[1])

x = x_x
y = y_y
```

```
fig, ax = plt.subplots()
ax.barh(x, y, color="deepskyblue")
labels = ax.get_xticklabels()
plt.setp(labels, rotation=0, horizontalalignment='right')

for a, b in zip(x, y):
    plt.text(b+1, a, b, ha='center', va='center')
ax.legend(["label"],loc="lower right")

plt.rcParams['font.sans-serif'] = ['SimHei']  # 用来正常显示中文标签
plt.ylabel('name')
plt.xlabel('数量')
plt.rcParams['savefig.dpi'] = 300  # 图片像素
plt.rcParams['figure.dpi'] = 300  # 分辨率
plt.rcParams['figure.figsize'] = (15.0, 8.0)  # 尺寸
plt.title("title")

plt.savefig('result.png')
plt.show()
```

执行后的效果如图 4-15 所示。

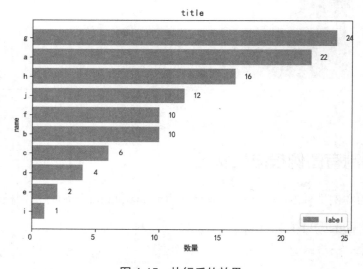

图 4-15　执行后的效果

4.1.11　绘制分组柱状图

请看下面的实例文件 bar11.py，功能是使用库 matplotlib 绘制有图例的横向分组柱状图。

 源码路径：**codes\4\4-1\bar11.py**

```
import matplotlib.pyplot as plt
import numpy as np

# 重复固定的随机状态
np.random.seed(19680801)
```

```python
plt.rcdefaults()
fig, ax = plt.subplots()

# 示例数据
people = ('Tom', 'Dick', 'Harry', 'Slim', 'Jim')
y_pos = np.arange(len(people))
performance = 3 + 10 * np.random.rand(len(people))
performance2 = 3 + 10 * np.random.rand(len(people))
error = np.random.rand(len(people))

total_width, n = 0.8, 2
width = total_width / n
y_pos=y_pos - (total_width - width) / 2

b=ax.barh(y_pos, performance, align='center',
        color='green', ecolor='black',height=0.2,label='a')
#添加数据标签
for rect in b:
    w=rect.get_width()
    ax.text(w,rect.get_y()+rect.get_height()/2,'%f'%w,ha='left',va='center')

b=ax.barh(y_pos+width, performance2, align='center',
        color='red', ecolor='black',height=0.2,label='b')
#添加数据标签
for rect in b:
    w=rect.get_width()
    ax.text(w,rect.get_y()+rect.get_height()/2,'%f'%w,ha='left',va='center')
ax.set_yticks(y_pos+width/2.0)
ax.set_yticklabels(people)
ax.invert_yaxis()  # labels read top-to-bottom
ax.set_xlabel('Performance')
ax.set_title('How fast do you want to go today?')
plt.legend()
plt.show()
print(y_pos+3)
```

执行后的效果如图 4-16 所示。

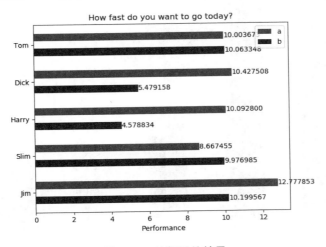

图 4-16　执行后的效果

4.1.12 模拟电影票房柱状图

请看下面的实例文件 bar12.py，功能是使用库 matplotlib 绘制模拟电影票房柱状图。

源码路径：codes\4\4-1\bar12.py

```python
import matplotlib.pyplot as plt
import numpy as np

# 设置matplotlib正常显示中文和负号
plt.rcParams['font.sans-serif'] = ['SimHei']
plt.rcParams['axes.unicode_minus'] = False

# 生成画布
plt.figure(figsize=(20, 8), dpi=80)
# 横坐标电影名字
movie_name = ['雷神3：诸神黄昏', '正义联盟', '东方快车谋杀案', '寻梦环游记', '全球风暴', '降魔传', '追捕', '七十七天', '密战', '狂兽', '其他']
# 纵坐标票房
y = [73853, 57767, 22354, 15969, 14839, 8725, 8716, 8318, 7916, 6764, 52222]
x = range(len(movie_name))

plt.bar(x, y, width=0.5, color=['b', 'r', 'g', 'y', 'c', 'm', 'y', 'k', 'c', 'g', 'g'])
plt.xticks(x, movie_name)

plt.show()
```

执行后的效果如图 4-17 所示。

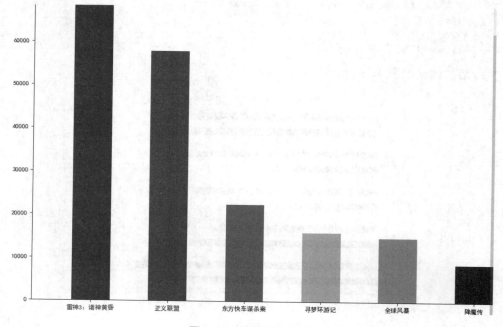

图 4-17　执行后的效果

4.1.13 绘制正负柱状图

请看下面的实例文件 bar13.py，功能是使用库 matplotlib 绘制正负柱状图。

源码路径：codes\4\4-1\bar13.py

```
import numpy as np
import matplotlib.pyplot as plt

a = np.array([5, 20, 15, 25, 10])
b = np.array([10, 15, 20, 15, 5])
plt.barh(range(len(a)), a)
plt.barh(range(len(b)), -b)
plt.show()
```

执行后的效果如图 4-18 所示。

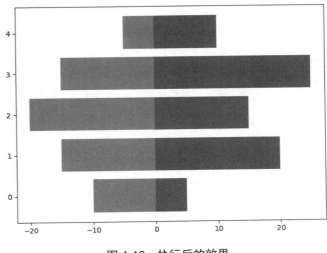

图 4-18 执行后的效果

4.1.14 绘制不同商品销量的统计柱状图

请看下面的实例文件 bar14.py，功能是使用库 matplotlib 绘制不同商品销量的统计柱状图。

源码路径：codes\4\4-1\bar14.py

```
import matplotlib.pyplot as plt

x=[1,2,3,4,5,6,7,8,9,10,11,12]
y=[6,3,9,2,6,16,8,10,4,14,18,6]

def get_color(x, y):
    """对销量不同的区段标为不同的颜色"""
    color = []
    for i in range(len(x)):
        if y[i] < 5:
```

```
            color.append("green")
        elif y[i] < 10:
            color.append("lightseagreen")
        elif y[i] < 15:
            color.append("gold")
        else:
            color.append("coral")

    return color

plt.bar(x,y,label="销量",color=get_color(x,y), tick_label=x)

for a,b in zip(x, y):
    plt.text(a, b+0.1, b, ha='center', va='bottom')

plt.legend(loc="upper left")
plt.rcParams['font.sans-serif'] = ['SimHei']   # 用来正常显示中文标签
plt.ylabel('销量')
plt.xlabel('date')
plt.rcParams['savefig.dpi'] = 300    # 图片像素
plt.rcParams['figure.dpi'] = 300     # 分辨率
plt.rcParams['figure.figsize'] = (15.0, 8.0)  # 尺寸
plt.title("月份销量的分布情况")
plt.savefig('D:\\result.png')
plt.show()
```

执行后的效果如图 4-19 所示。

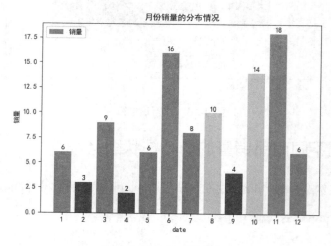

图 4-19　执行后的效果

4.2　可视化分析掷骰子游戏的结果次数

骰子也叫色子，是用象牙、骨头或塑料做的小正方体块。每面刻有点数，一到六，常用于做各种游戏。在掷骰子时先摇动骰子，

扫码观看本节视频讲解

然后抛掷使两个骰子都随意停止在一平面上。在本实例中将模拟掷骰子游戏，并统计两个骰子的结果点数，可视化分析不同结果出现的次数。

4.2.1 使用库 pygal 实现模拟掷骰子功能

在下面的实例文件 01.py 中，演示了使用 pygal 库实现模拟掷骰子功能的过程。首先定义了骰子类 Die，然后使用函数 range( ) 模拟掷骰子 1000 次，之后统计每个骰子点数的出现次数，最后在柱形图中显示统计结果。文件 01.py 的具体实现代码如下。

源码路径：codes\4\4-2\01.py

```python
import random
class Die:
    """
    一个骰子类
    """
    def __init__(self, num_sides=6):
        self.num_sides = num_sides

    def roll(self):
        return random.randint(1, self.num_sides)

import pygal

die = Die()
result_list = []
# 掷1000次
for roll_num in range(1000):
    result = die.roll()
    result_list.append(result)

frequencies = []
# 范围1~6,统计每个数字出现的次数
for value in range(1, die.num_sides + 1):
    frequency = result_list.count(value)
    frequencies.append(frequency)

# 条形图
hist = pygal.Bar()
hist.title = 'Results of rolling one D6 1000 times'
# x轴坐标
hist.x_labels = [1, 2, 3, 4, 5, 6]
# x、y轴的描述
hist.x_title = 'Result'
hist.y_title = 'Frequency of Result'
# 添加数据,第一个参数是数据的标题
hist.add('D6', frequencies)
# 保存到本地,格式必须是svg
hist.render_to_file('die_visual.svg')
```

执行后会生成一个名为 die_visual.svg 的文件，可以用浏览器打开这个 SVG 文件，打开后会显示统计柱形图。执行效果如图 4-20 所示。如果将光标指向数据,可以看到显示了标题 D6、x 轴的坐标以及 y 轴坐标。6 个数字出现的频次是差不多的,其实理论上概率是 1/6,随着试验次数的增加,趋势会变得越来越明显。

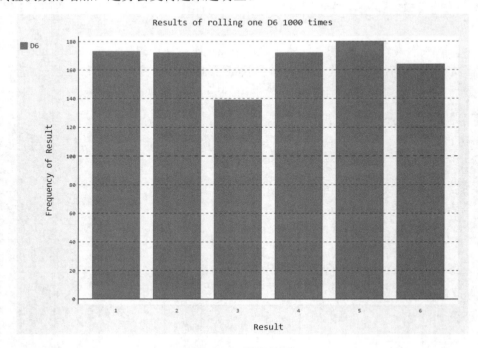

图 4-20　执行效果

4.2.2　同时掷两个骰子

可以对上面的实例进行升级,如同时掷两个骰子,可以通过下面的实例文件 02.py 实现。在具体实现时,首先定义骰子类 Die,然后使用函数 range( )模拟掷两个骰子 5000 次,之后统计每次掷两个骰子点数的最大数的次数,最后在柱形图中显示统计结果。文件 02.py 的具体实现代码如下。

源码路径:**codes\4\4-2\02.py**

```
class Die:
    """
    一个骰子类
    """
    def __init__(self, num_sides=6):
        self.num_sides = num_sides

    def roll(self):
        return random.randint(1, self.num_sides)
die_1 = Die()
die_2 = Die()

result_list = []
```

```
for roll_num in range(5000):
    # 两个骰子的点数和
    result = die_1.roll() + die_2.roll()
    result_list.append(result)

frequencies = []
# 能掷出的最大数
max_result = die_1.num_sides + die_2.num_sides

for value in range(2, max_result + 1):
    frequency = result_list.count(value)
    frequencies.append(frequency)

# 可视化
hist = pygal.Bar()
hist.title = 'Results of rolling two D6 dice 5000 times'
hist.x_labels = [x for x in range(2, max_result + 1)]
hist.x_title = 'Result'
hist.y_title = 'Frequency of Result'
# 添加数据
hist.add('two D6', frequencies)
# 格式必须是svg
hist.render_to_file('2_die_visual.svg')
```

执行后会生成一个名为 2_die_visual.svg 的文件，可以用浏览器打开这个 SVG 文件，打开后会显示统计柱形图。执行效果如图 4-21 所示。由此可以看出，两个骰子之和为 7 的次数最多，和为 2 的次数最少。因为能掷出 2 的只有一种情况(1, 1)；而掷出 7 的情况有(1, 6) (2, 5) (3, 4) (4, 3) (5, 2) (6, 1)共 6 种情况，其余数字的情况都没有 7 的多，故掷得 7 的概率最大。

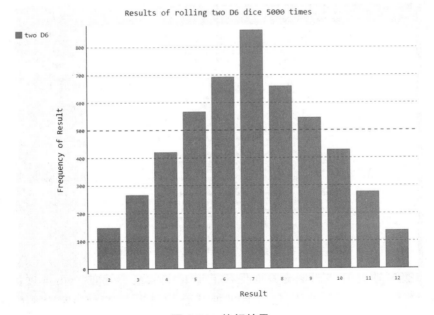

图 4-21　执行效果

4.3 可视化分析最受欢迎的开源项目

在现实应用中，经常需要可视化分析网络数据。例如，想分析某开源网站中最受欢迎的 Python 库，要求以 stars 进行排序，应该如何实现呢？本节将讲解实现这一功能的方法。

4.3.1 统计前 30 名最受欢迎的 Python 库

扫码观看本节视频讲解

对于广大开发者来说，×××网站是大家心中的殿堂，在里面有无数个开源程序供开发者学习和使用。为了便于开发者了解×××网站中每个项目的基本信息，×××网站官方提供了一个 JSON 网页，在里面存储了按照某个标准排列的项目信息，如通过以下网址可以查看关键字是 python、按照 stars 从高到低排列的项目信息，如图 4-22 所示。

```
https://api.XXX网站.com/search/repositories?q=language:python
```

图 4-22　按照 stars 从高到低排列的 Python 项目

对上述 JSON 数据中，在 items 里面保存了前 30 名 stars 最多的 Python 项目信息。其中 name 表示库名称，owner 下的 login 是库的拥有者，html_url 表示该库的网址(注意 owner 下也有个 html_url，但它是用户的×××网站网址，要定位到该用户的具体这个库，所以不要用 owner 下的 html_url，stargazers_count 表示所得的 stars 数目。

另外，total_count 表示 Python 语言的仓库的总数。incomplete_results 表示响应的值是否不完全，一般来说是 false，表示响应的数据完整。

在下面的实例文件 github01.py 中，演示了使用 requests 获取网站中前 30 名最受欢迎的 Python 库信息的方法：

```
import requests
url = 'https://api.github.com/search/repositories?q=language:python&sort=stars'
response = requests.get(url)
# 200 为响应成功
print(response.status_code, '响应成功！')
response_dict = response.json()

total_repo = response_dict['total_count']
repo_list = response_dict['items']
print('总仓库数: ', total_repo)
print('top', len(repo_list))
for repo_dict in repo_list:
    print('\n 名字: ', repo_dict['name'])
    print('作者: ', repo_dict['owner']['login'])
    print('Stars: ', repo_dict['stargazers_count'])
    print('网址: ', repo_dict['html_url'])
    print('简介: ', repo_dict['description'])
```

执行后会提取 JSON 数据中的信息，输出显示×××网站中前 30 名最受欢迎的 Python 库信息：

```
200 响应成功！
总仓库数: 4965355
top 30

名字:  system-design-primer
作者:  donnemartin
Stars:  95031
网址:  https://github.com/donnemartin/system-design-primer
简介:  Learn how to design large-scale systems. Prep for the system design interview. Includes Anki flashcards.

名字:  awesome-python
作者:  vinta
Stars:  82369
网址:  https://github.com/vinta/awesome-python
简介:  A curated list of awesome Python frameworks, libraries, software and resources.

名字:  Python
作者:  TheAlgorithms
Stars:  71386
网址:  https://github.com/TheAlgorithms/Python
简介:  All Algorithms implemented in Python.
```

```
名字： youtube-dl
作者： ytdl-org
Stars： 66053
网址： https://github.com/ytdl-org/youtube-dl
简介： Command-line program to download videos from YouTube.com and other video
sites.

名字： models
作者： tensorflow
Stars： 63658
网址： https://github.com/tensorflow/models
简介： Models and examples built with TensorFlow.

名字： thefuck
作者： nvbn
Stars： 53445
网址： https://github.com/nvbn/thefuck
简介： Magnificent app which corrects your previous console command.

名字： flask
作者： pallets
Stars： 50287
网址： https://github.com/pallets/flask
简介： The Python micro framework for building web applications.

名字： django
作者： django
Stars： 49243
网址： https://github.com/django/django
简介： The Web framework for perfectionists with deadlines.

名字： keras
作者： keras-team
Stars： 48182
网址： https://github.com/keras-team/keras
简介： Deep Learning for humans.

########在后面省略其余的结果
```

4.3.2 使用 pygal 实现数据可视化

虽然通过实例文件 github01.py 可以提取 JSON 页面中的数据，但是数据还不够直观，接下来编写实例文件 github02.py，将从 github 网站的总仓库中提取最受欢迎的 Python 库(前 30 名)，并绘制统计直方图。文件 github02.py 的具体实现代码如下：

```
import requests

import pygal
from pygal.style import LightColorizedStyle, LightenStyle
```

```python
url = 'https://api.github.com/search/repositories?q=language:python&sort=stars'
response = requests.get(url)
# 200 为响应成功
print(response.status_code, '响应成功!')
response_dict = response.json()

total_repo = response_dict['total_count']
repo_list = response_dict['items']
print('总仓库数: ', total_repo)
print('top', len(repo_list))

names, plot_dicts = [], []
for repo_dict in repo_list:
    names.append(repo_dict['name'])
    # 加上 str 强转,否则会遇到 'NoneType' object is not subscriptable 错误
    plot_dict = {
        'value' : repo_dict['stargazers_count'],
        # 有些描述很长,选最前一部分
        'label' : str(repo_dict['description'])[:200]+'...',
        'xlink' : repo_dict['html_url']
    }
    plot_dicts.append(plot_dict)

# 改变默认主题颜色,偏蓝色
my_style = LightenStyle('#333366', base_style=LightColorizedStyle)
# 配置
my_config = pygal.Config()
# x 轴的文字旋转 45 度
my_config.x_label_rotation = -45
# 隐藏左上角的图例
my_config.show_legend = False
# 标题字体大小
my_config.title_font_size = 30
# 副标签,包括 x 轴和 y 轴大部分
my_config.label_font_size = 20
# 主标签是 y 轴某数倍数,相当于一个特殊的刻度,让关键数据点更醒目
my_config.major_label_font_size = 24
# 限制字符为 15 个,超出的以...显示
my_config.truncate_label = 15
# 不显示 y 参考虚线
my_config.show_y_guides = False
# 图表宽度
my_config.width = 1000

# 第一个参数可以传配置
chart = pygal.Bar(my_config, style=my_style)
chart.title = '×××网站最受欢迎的 Python 库(前 30 名)'
# x 轴的数据
chart.x_labels = names
# 加入 y 轴的数据,无须 title 设置为空,注意这里传入的字典,
# 其中的键 value 是 y 轴的坐标值
chart.add('', plot_dicts)
chart.render_to_file('30_stars_python_repo.svg')
```

执行后会创建生成数据统计直方图文件 30_stars_python_repo.svg，并输出以下提取信息：

```
200 响应成功！
总仓库数：3394860
top 30
```

数据可视化统计柱状图文件 30_stars_python_repo.svg 的执行效果如图 4-23 所示。

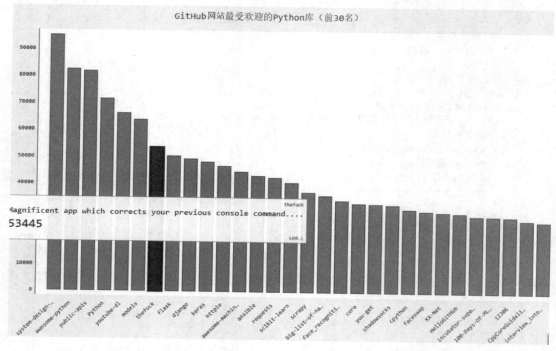

图 4-23　统计柱状图文件 30_stars_python_repo.svg 的执行效果

扫码观看本节视频讲解

4.4　可视化统计显示某网店各类口罩的销量

假设有一个在天猫平台销售口罩的网店，在一个 CSV 文件中保存了 2020 年第一季度各类口罩的销量。可以编写一个 Python 程序，使用 Matplotlib 绘制可视化各类口罩销量的统计柱状图。

4.4.1 准备 CSV 文件

准备 CSV 文件 fall-2020.csv，在里面保存了各类口罩的销量数据，文件 fall-2020.csv 的具体内容如下：

```
7
普通口罩,1244
普通医用口罩,1142
医用外科口罩,4250
N90 口罩,1754
N95 口罩,1569
KN94 口罩,3763
N99 口罩,1149
```

4.4.2 可视化 CSV 文件中的数据

编写 Python 程序文件 n95.py，功能是读取 CSV 文件 fall-2020.csv 的内容，并根据读取的数据绘制柱状图，展示一个类口罩商品的销量情况。程序文件 n95.py 的具体实现代码如下：

```python
import numpy as np
import matplotlib.pyplot as plt
plt.rcParams['font.sans-serif'] = ['SimHei']  # 指定默认字体

def read_file(file_name):  #读取CSV文件的内容
    i = 0
    data = open(file_name, 'r', encoding='UTF-8')
    for line in data:
        sections = int(line[0])
        college_array = np.empty(sections, dtype = object)  #为不同口罩类型创建空数组
        enrollment_array = np.zeros(sections)  #为销量创建空数组
        break

    #使用逗号分隔数据,然后使用for循环将其添加到分隔数组中
    for line in data:
        line = line.split(',')
        line[1] = line[1][:-1]
        college_names = line[0]
        college_array[i] = college_names
        enrollment_array[i] = line[1]
        i +=1
    return college_array, enrollment_array  #分别返回college department 和 statistic 数组的元组

def main(file_name):
```

```
    i = 0
    read_file(file_name)  #读取 CSV 文件的内容
    college_names, college_enrollments = read_file(file_name) #gets the results
from the read_file method
    print(college_names)
    print(college_enrollments)

    #使用 matplotlib 库绘制图形并以可读的方式显示信息
    plt.figure("2020 某网店的口罩销量统计图")
    plt.bar(college_names, college_enrollments, width = 0.8, color = ["blue",
"gold"])
    plt.yticks(ticks = np.arange(0,4800, 400))
    plt.ylim(0,4400)

    plt.xlabel("口罩类型")
    plt.ylabel("销量")

    plt.show()#显示可视化统计图形
main("fall-2020.csv")
```

执行后的效果如图 4-24 所示。

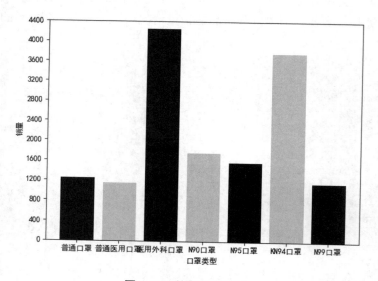

图 4-24　执行后的效果

4.5　数据挖掘：可视化处理文本情感分析数据

情感分析是自然语言处理和数据挖掘领域中的重要内容，也是机器学习开发者们乐于研究的领域。文本情感分析又称为意见挖掘、倾向性分析等，是对带有情感色彩的主观性文本进行分析、处理、归纳和推理的过程。在互联网(如博客和论坛以及社会服务网络如大

扫码观看本节视频讲解

众点评)中产生了大量的用户参与的、对于如人物、事件、产品等有价值的评论信息。可以编写一个 Python 程序，使用库 matplotlib 绘制可视化处理文本情感分类数据的柱状图。

4.5.1 准备 CSV 文件

预先准备两部电视剧的剧本文件，这些文件的格式是记事本格式.txt，在这些剧本文件中有大量的英文单词。将整个剧本文件分为两类，即 a 和 b，分别表示电视剧 a 和电视剧 b 的剧本文件。在 CSV 文件 sentiment_lex.csv 中保存了 4798 个情感分析单词的值，如图 4-25 所示。

图 4-25 准备的 CSV 文件

4.5.2 可视化两个剧本的情感分析数据

编写 Python 程序文件 main.py，具体实现流程如下。

(1) 打开需要的库，设置 font.sans-serif 保证能够在绘制的柱状图中显示中文。对应的实现代码如下：

```
import numpy as np
import matplotlib.pyplot as mplot
from glob import glob
mplot.rcParams['font.sans-serif'] = ['SimHei'] # 指定默认字体
t=0
t2=0
```

(2) 准备要处理的文件，设置第 1 部电视剧剧本对应的记事本文件名以字母 a 开头，第 2 部电视剧剧本对应的记事本文件名以字母 b 开头。对应的实现代码如下：

```
afilenames = glob('a*.txt')
allawords=[]

bfilenames = glob('b*.txt')
allbwords=[]

file=[]
```

```
lex={}
```

(3) 设置要在柱状图中统计显示的 5 种情感类型,即 neg(Negative,阴性)、wneg(Weakly Negative,弱阴性)、neu(Neutral,中性)、wpos(Weakly Positive,弱阳性)、pos(Positive,阳性)。对应的实现代码如下:

```
neg=0
wneg=0
neu=0
wpos=0
pos=0
```

(4) 遍历所有第 1 部电视剧剧本对应的记事本文件,分割每个单词的内容。对应的实现代码如下:

```
for i in range(0,len(afilenames)):
    data=open(afilenames[i], "r")
    data=data.read()
    data=data.replace("\n", " ")
    data=data.replace(".", "")
    data=data.replace("?", "")
    data=data.replace(",", "")
    data=data.replace("!", "")
    data=data.replace("[", "")
    data=data.replace("]", "")
    data=data.replace("(", "")
    data=data.replace(")", "")
    data=data.replace(">", "")
    data=data.replace("<", "")
    data=data.split(" ")
    for j in range(0, len(data)):
        if len(data[j]) > 0:
            if (data[j]) != (data[j].upper()):
                allawords.append(data[j].lower())
```

(5) 遍历所有第 2 部电视剧剧本对应的记事本文件,分割每个单词的内容。对应的实现代码如下:

```
for i in range(0,len(bfilenames)):
    data=open(bfilenames[i], "r")
    data=data.read()
    data=data.replace("\n", " ")
    data=data.replace(".", "")
    data=data.replace("?", "")
    data=data.replace(",", "")
    data=data.replace("!", "")
    data=data.replace("[", "")
    data=data.replace("]", "")
    data=data.replace("(", "")
    data=data.replace(")", "")
    data=data.split(" ")
    for j in range(0, len(data)):
        if len(data[j]) > 0:
            if (data[j]) != (data[j].upper()):
                allbwords.append(data[j].lower())
```

(6) 打开预先设置的 CSV 文件 sentiment_lex.csv,然后读取里面的情感分析值。对应的实现代码如下:

```
file=open("sentiment_lex.csv", "r")
sent=file.read()
sent=sent.split("\n")


for i in range(0,len(sent)-2):
    sent[i]=sent[i].split(",")
    sent[i][1]= np.float64(sent[i][1])
    lex[sent[i][0]]=sent[i][1]
```

(7) 提示用户输入要分析哪一部电视剧剧本,如果用户输入 a,则分析第 1 部电视剧的剧本内容,并使用库 matplotlib 绘制情感分析统计柱状图。对应的实现代码如下:

```
choice = input("分析哪一部电视剧剧本? ")
if choice == "a":
    for i in range(0, len(allawords)):
        if allawords[i] in lex:
            if lex[allawords[i]]>=-1 and lex[allawords[i]]<-.6:
                neg+=1
            if lex[allawords[i]]>=-.6 and lex[allawords[i]]<-.2:
                wneg+=1
            if lex[allawords[i]]>=-.2 and lex[allawords[i]]<=.2:
                neu+=1
            if lex[allawords[i]]>.2 and lex[allawords[i]]<=.6:
                wpos+=1
            if lex[allawords[i]]>.6 and lex[allawords[i]]<=1:
                pos+=1

    y = [neg, wneg, neu, wpos, pos]
    y=np.log10(y)
    x = ["Neg", "W.Neg", "Neu", "W.Pos", "Pos"]

    mplot.title("a 类情感分析")
    mplot.xlabel("情感")
    mplot.ylabel("log10")
    mplot.bar(x,y)
    mplot.show()
```

(8) 如果用户输入 b,则分析第 2 部电视剧剧本内容,并使用库 matplotlib 绘制情感分析统计柱状图。对应的实现代码如下:

```
elif choice == "b":
    for i in range(0, len(allbwords)):
        if allbwords[i] in lex:
            if lex[allbwords[i]]>=-1 and lex[allbwords[i]]<-.6:
                neg+=1
            if lex[allbwords[i]]>=-.6 and lex[allbwords[i]]<-.2:
                wneg+=1
            if lex[allbwords[i]]>=-.2 and lex[allbwords[i]]<=.2:
                neu+=1
            if lex[allbwords[i]]>.2 and lex[allbwords[i]]<=.6:
```

```
                wpos+=1
            if lex[allbwords[i]]>.6 and lex[allbwords[i]]<=1:
                pos+=1


y = [neg, wneg, neu, wpos, pos]
y=np.log10(y)
x = ["Neg", "W.Neg", "Neu", "W.Pos", "Pos"]

mplot.title("b类情感分析")
mplot.xlabel("情感")
mplot.ylabel("log10")
mplot.bar(x,y)
mplot.show()
```

执行后会提示用户输入要分析哪一部电视剧剧本,如果用户输入 a,则分析第 1 部电视剧剧本内容,并使用库 matplotlib 绘制情感分析统计柱状图,如图 4-26 所示。

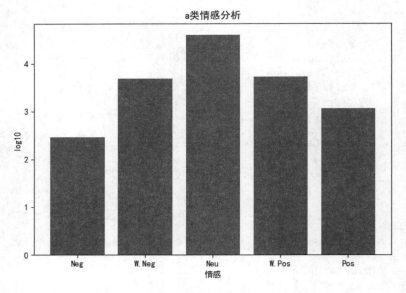

图 4-26　绘制的情感分析统计柱状图

第 5 章

绘制饼状图

饼状图常用于数据统计和分析领域,通常分为二维与三维饼状图。饼状图显示一个数据系列(数据系列:在图表中绘制的相关数据点,这些数据源自数据表的行或列。在现实应用中,经常使用饼状图来展示数据分析的结果,这样可以更加直观地展示数据分析结果。本章将详细讲解使用 Python 绘制饼状图的知识。

5.1 绘制基本的饼状图

在数据可视化分析工作中，经常需要进行数据统计并绘制相关的饼状图。本节将详细讲解在 Python 程序中使用各种库绘制基本饼状图的知识。

5.1.1 绘制简易的饼状图

扫码观看本节视频讲解

在 Python 程序中，可以使用 Matplotlib 中的函数 matplotlib.pyplot.pie( )绘制饼状图。请看下面的实例文件 pie01.py，功能是使用 matplotlib 库绘制一个饼状图，可视化展示××网站会员用户教育水平的分布信息。

```
import matplotlib.pyplot as plt
plt.rcParams['font.sans-serif'] = ['SimHei'] # 指定默认字体
# 构造数据
edu = [0.2515,0.3724,0.3336,0.0368,0.0057]
labels = ['中专','大专','本科','硕士','其他']
"""
# 绘制饼图
explode: 设置各部分突出
label:设置各部分标签
labeldistance:设置标签文本距圆心位置,1.1 表示1.1 倍半径
autopct: 设置圆里面文本
shadow: 设置是否有阴影
startangle: 起始角度，默认从 0 开始逆时针方向旋转
pctdistance: 设置圆内文本距圆心距离
返回值
l_text: 圆内部文本,matplotlib.text.Text object
p_text: 圆外部文本
"""

# 绘制饼图
plt.pie(x = edu, # 绘图数据
    labels=labels, # 添加教育水平标签
    autopct='%.1f%%' # 设置百分比的格式,这里保留一位小数
)
# 添加图标题
plt.title('XX 网站会员用户的教育水平分布')
plt.show()                    # 显示图形
```

执行后的效果如图 5-1 所示。

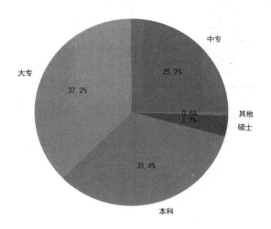

图 5-1 执行后的效果

另外，也可以使用库 pygal 中内置函数 pie( )绘制饼状图。请看下面的实例文件 pie01-1.py，功能是库 pygal 绘制一个饼状图，可视化展示某年度浏览器产品的市场份额数据。

```
import pygal

pie_chart = pygal.Pie()
pie_chart.title = 'Browser usage in February 2012 (in %)'
pie_chart.add('IE', 19.5)
pie_chart.add('Firefox', 36.6)
pie_chart.add('Chrome', 36.3)
pie_chart.add('Safari', 4.5)
pie_chart.add('Opera', 2.3)
pie_chart.render_to_file('bar_chart.svg')
```

执行后会得到文件 bar_chart.svg，效果如图 5-2 所示。

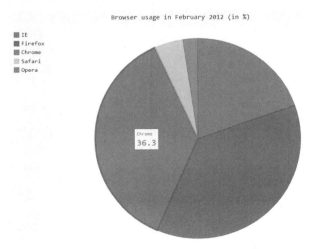

图 5-2 文件 bar_chart.svg 的执行效果

5.1.2 修饰饼状图

请看下面的实例文件 pie02.py,功能是以前面的实例文件 pie01.py 为基础进行升级,修饰了使用库 matplotlib 绘制的饼状图,可视化展示××网站会员用户教育水平的分布信息。

```python
import matplotlib.pyplot as plt

plt.rcParams['font.sans-serif'] = ['SimHei']  # 指定默认字体
edu = [0.2515,0.3724,0.3336,0.0368,0.0057]
labels = ['中专','大专','本科','硕士','其他']
# 添加修饰的饼图
explode = [0,0.1,0,0,0]# 生成数据,用于突出显示大专学历人群
colors=['#9999ff','#ff9999','#7777aa','#2442aa','#dd5555']# 自定义颜色

# 中文乱码和坐标轴负号的处理
plt.rcParams['font.sans-serif'] = ['Microsoft YaHei']
plt.rcParams['axes.unicode_minus'] = False

# 将横、纵坐标轴标准化处理,确保饼图是一个正圆;否则为椭圆
plt.axes(aspect='equal')
# 绘制饼图
plt.pie(x = edu,                # 绘图数据
    explode=explode,            # 突出显示大专人群
    labels=labels,              # 添加教育水平标签
    colors=colors,              # 设置饼图的自定义填充色
    autopct='%.1f%%',           # 设置百分比的格式,这里保留一位小数
    pctdistance=0.8,            # 设置百分比标签与圆心的距离
    labeldistance = 1.1,        # 设置教育水平标签与圆心的距离
    startangle = 180,           # 设置饼图的初始角度
    radius = 1.2,               # 设置饼图的半径
    counterclock = False,       # 是否逆时针方向旋转,这里设置为顺时针方向旋转
    wedgeprops = {'linewidth': 1.5, 'edgecolor':'green'},# 设置饼图内外边界的属性值
    textprops = {'fontsize':10, 'color':'black'}, # 设置文本标签的属性值
)

# 添加图标题
plt.title('XX 网站会员用户的教育水平分布统计图')
# 显示图形
plt.show()
```

执行后的效果如图 5-3 所示。

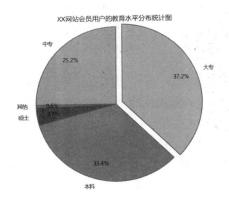

图 5-3 执行后的效果

5.1.3 突出显示某个饼状图的部分

在前面的实例文件 pie02.py 中，通过设置函数 matplotlib.pyplot.pie( )中的参数 explode 设置某个饼状部分突出显示。请看下面的实例文件 pie03.py，功能是使用 matplotlib 绘制的饼状图，设置了某个饼状图部分突出显示。

```
import matplotlib.pyplot as plt
plt.rcParams['font.sans-serif']='SimHei'         #设置中文显示
plt.figure(figsize=(6,6))                         #将画布设定为正方形,则绘制的饼图是正圆
label=['第一','第二','第三']                      #定义饼图的标签,标签是列表
explode=[0.01,0.2,0.01]                           #设定各项距离圆心 n 个半径
#plt.pie(values[-1,3:6],explode=explode,labels=label,autopct='%1.1f%%')#绘制饼图
values=[4,7,9]
plt.pie(values,explode=explode,labels=label,autopct='%1.1f%%')#绘制饼图
plt.title('2018年饼图')
plt.savefig('./2018年饼图')
plt.show()
```

执行后的效果如图 5-4 所示。

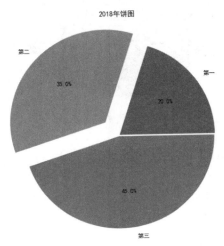

图 5-4 执行后的效果

5.1.4 为饼状图添加图例

在使用 matplotlib 绘制饼状图时，可以通过内置函数 legend( )为绘制饼状图添加图例。请看下面的实例文件 pie04.py，功能是使用 matplotlib 库绘制饼状图，并为绘制饼状图添加图例说明。

```
import matplotlib.pyplot as plt
import matplotlib

matplotlib.rcParams['font.sans-serif'] = ['SimHei']
matplotlib.rcParams['axes.unicode_minus'] = False

label_list = ["第一部分", "第二部分", "第三部分"]    # 各部分标签
size = [55, 35, 10]    # 各部分大小
color = ["red", "green", "blue"]    # 各部分颜色
explode = [0.05, 0, 0]    # 各部分突出值

patches, l_text, p_text = plt.pie(size, explode=explode, colors=color,
labels=label_list, labeldistance=1.1, autopct="%1.1f%%", shadow=False,
startangle=90, pctdistance=0.6)
plt.axis("equal")    # 设置横轴和纵轴大小相等，这样饼才是圆的
plt.legend()
plt.show()
```

执行后的效果如图 5-5 所示。

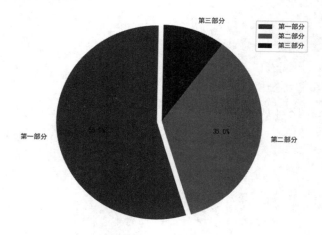

图 5-5　执行后的效果

5.1.5 使用饼状图可视化展示某地区程序员的工龄

请看下面的实例文件 pie05.py，功能是使用库 matplotlib 绘制的饼状图，可视化展示某地区程序员的工龄分部信息。

```
import matplotlib.pyplot as plt
from matplotlib.font_manager import FontProperties
```

```python
font = FontProperties(fname=r"C:\Windows\Fonts\simhei.ttf", size=14)
beijing = [17,17,23,43]
shanghai = ['19%','4%','23%','54%']
guangzhou = ['53%','25%','13%','9%']
shenzhen = ['41%','22%','20%','17%']
label = ['2-3 years','3-4 years','4-5 years','5+ years']
color = ['red','green','yellow','purple']
indic = []

#将数据最大的突出显示
for value in beijing:
    if value == max(beijing):
        indic.append(0.1)
    else:
        indic.append(0)

plt.pie(
    beijing,
    labels=label,
    colors=color,
    startangle=90,
    shadow=True,
    explode=tuple(indic),#tuple方法用于将列表转化为元组
    autopct='%1.1f%%'#是数字1,不是l
)

plt.title(u'统计XX地区程序员的工龄', FontProperties=font)

plt.show()
```

执行后的效果如图 5-6 所示。

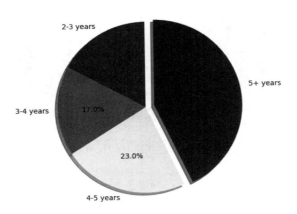

图 5-6 执行后的效果

5.1.6 绘制多个饼状图

(1) 请看下面的实例文件 pie06.py, 功能是使用 matplotlib 库同时绘制 4 个饼状图。

```python
import matplotlib.pyplot as plt
import matplotlib.font_manager as fm
myfont = fm.FontProperties(fname=r'C:\Windows\Fonts\simhei.ttf')
fig = plt.figure("tu")
ax_list = []
labels_list= [['娱乐','育儿','饮食','房贷','交通','其它',"坚持"],['Sur', "Fea",
"Dis", "Hap", "Sad", "Ang", "Nat"]]
sizes_list= [[2,5,12,70,2,2,7]]
explode_list=[ (0,0,0,0.1,0,0,0)]
title_list=["头部左右","头部上下","头部垂面旋转","表情"]
for ax_one in range(4):
    ax1 = fig.add_subplot(2,2, ax_one + 1)

ax1.pie(sizes_list[0],explode=explode_list[0],labels=labels_list[1],autopct
='%1.1f%%',shadow=False,startangle=150)
    plt.title(title_list[ax_one],fontproperties=myfont)

plt.show()
if __name__ == '__main__':
    pass
```

执行后的效果如图 5-7 所示。

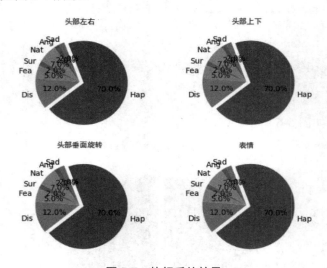

图 5-7 执行后的效果

(2) 请看下面的实例文件 pie07.py，功能是使用 matplotlib 库同时绘制 3 个并排的饼状图。

```
import matplotlib.pyplot as plt

labels = 'Comments rated 1', 'Comments rated 2', 'Comments rated 3', 'Comments
rated 4', 'Comments rated 5'
sizes = [38, 16, 18, 54, 107]
sizes1 = [84, 45, 59, 132, 314]
sizes2 = [205, 172, 273, 561, 1954]
explode = (0, 0, 0, 0, 0.025)   # 设置各部分距离圆心的距离
fig1 = plt.figure(facecolor='white',figsize=(16,8))
```

```
ax1=plt.subplot(1,3,1)
ax1.pie(sizes, explode=explode,autopct='%1.1f%%',
        shadow=False, startangle=90)
ax1.axis('equal')
ax1.legend(labels)

ax1=plt.subplot(1,3,2)
ax1.pie(sizes1, explode=explode,autopct='%1.1f%%',
        shadow=False, startangle=90)
ax1.axis('equal')
ax1.legend(labels)

ax1=plt.subplot(1,3,3)
ax1.pie(sizes2, explode=explode,autopct='%1.1f%%',
        shadow=False, startangle=90)
ax1.axis('equal')
ax1.legend(labels)

plt.tight_layout()
plt.show()
fig1.savefig('饼状图.jpg',dpi=400)
```

执行后的效果如图 5-8 所示。

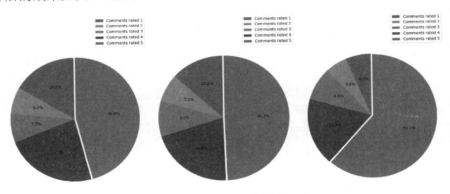

图 5-8 执行后的效果

(3) 请看下面的实例文件 pie08.py，功能是使用库 matplotlib 绘制一个 3 行 2 列的饼状图。

```
import numpy as np
import matplotlib.pyplot as plt
import pylab as pl
#画出3行2列的饼图

labels = ['Flat','Reduce','Raise']
# 321 > 3行2列第1个
fig1 = pl.subplot(321)
pl.pie([12,1,7],labels=labels,autopct='%1.1f%%',shadow=False,startangle=90)
plt.axis('equal')
plt.title("group A_20min")

# 322 > 3行2列第2个
```

```python
fig2 = pl.subplot(322)
pl.pie([2,2,1],labels=labels,autopct='%1.1f%%',shadow=False,startangle=90)
plt.axis('equal')
plt.title("group B_20min")

# 323 > 3行2列第3个
fig3 = pl.subplot(323)
pl.pie([8,2,10],labels=labels,autopct='%1.1f%%',shadow=False,startangle=90)
plt.axis('equal')
plt.title("group A_60min")

# 324 > 3行2列第4个
fig4 = pl.subplot(324)
pl.pie([2,1,2],labels=labels,autopct='%1.1f%%',shadow=False,startangle=90)
plt.axis('equal')
plt.title("group B_60min")

# 325 > 3行2列第5个
fig5 = pl.subplot(325)
pl.pie([4,1,15],labels=labels,autopct='%1.1f%%',shadow=False,startangle=90)
#startangle 表示饼图的起始角度
plt.axis('equal')  #这行代码加入饼图不会画成椭圆
plt.title("group A_70min")

# 326 > 3行2列第6个
fig6 = pl.subplot(326)
pl.pie([3,1,1],labels=labels,autopct='%1.1f%%',shadow=False,startangle=90)
plt.axis('equal')
plt.title("group B_70min")

pl.tight_layout() #布局方法
pl.savefig('vc5.jpg',dpi = 500) #dpi 实参改变图像的分辨率
pl.show() #显示方法
```

执行后的效果如图 5-9 所示。

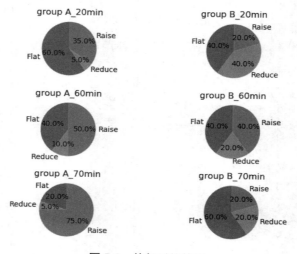

图 5-9　执行后的效果

5.1.7 绘制多系列饼状图

请看下面的实例文件 pie09.py，功能是使用库 pygal 绘制多系列饼状图，可视化展示某年度浏览器产品的市场份额数据。

```
import pygal

pie_chart = pygal.Pie()
pie_chart.title = 'Browser usage by version in February 2012 (in %)'
pie_chart.add('IE', [5.7, 10.2, 2.6, 1])
pie_chart.add('Firefox', [.6, 16.8, 7.4, 2.2, 1.2, 1, 1, 1.1, 4.3, 1])
pie_chart.add('Chrome', [.3, .9, 17.1, 15.3, .6, .5, 1.6])
pie_chart.add('Safari', [4.4, .1])
pie_chart.add('Opera', [.1, 1.6, .1, .5])
pie_chart.render_to_file('bar_chart.svg')
```

执行后会生成文件 bar_chart.svg，效果如图 5-10 所示。

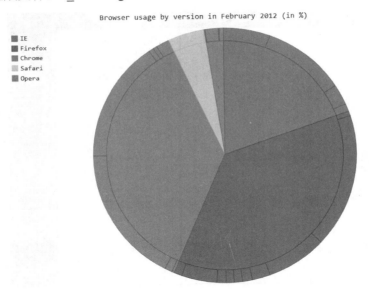

图 5-10　文件 bar_chart.svg 的执行效果

5.1.8 绘制圈状饼状图

请看下面的实例文件 pie10.py，功能是使用库 pygal 绘制圈状饼状图，可视化展示某年度浏览器产品的市场份额数据。

```
import pygal

pie_chart = pygal.Pie(inner_radius=.4)
pie_chart.title = 'Browser usage in February 2012 (in %)'
pie_chart.add('IE', 19.5)
pie_chart.add('Firefox', 36.6)
```

```
pie_chart.add('Chrome', 36.3)
pie_chart.add('Safari', 4.5)
pie_chart.add('Opera', 2.3)
pie_chart.render_to_file('bar_chart.svg')
```

执行后会生成文件 bar_chart.svg，效果如图 5-11 所示。

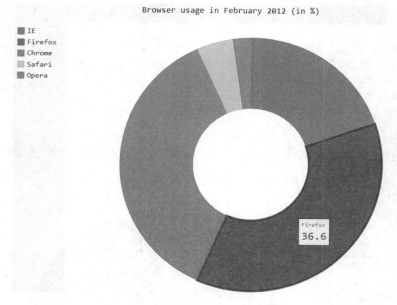

图 5-11 文件 bar_chart.svg 的执行效果

5.1.9 绘制环状饼状图

请看下面的实例文件 pie11.py，功能是使用库 pygal 绘制环状饼状图，可视化展示某年度浏览器产品的市场份额数据。

```
import pygal

pie_chart = pygal.Pie(inner_radius=.75)
pie_chart.title = 'Browser usage in February 2012 (in %)'
pie_chart.add('IE', 19.5)
pie_chart.add('Firefox', 36.6)
pie_chart.add('Chrome', 36.3)
pie_chart.add('Safari', 4.5)
pie_chart.add('Opera', 2.3)
pie_chart.render_to_file('bar_chart.svg')
```

执行后会生成文件 bar_chart.svg，效果如图 5-12 所示。

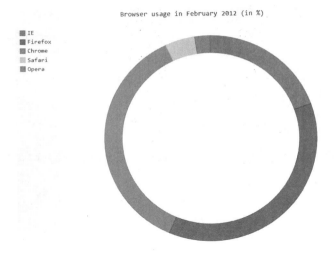

图 5-12　文件 bar_chart.svg 的执行效果

5.1.10　绘制半饼状图

请看下面的实例文件 pie12.py，功能是使用库 pygal 绘制半饼状图，可视化展示某年度浏览器产品的市场份额数据。

```
import pygal

pie_chart = pygal.Pie(half_pie=True)
pie_chart.title = 'Browser usage in February 2012 (in %)'
pie_chart.add('IE', 19.5)
pie_chart.add('Firefox', 36.6)
pie_chart.add('Chrome', 36.3)
pie_chart.add('Safari', 4.5)
pie_chart.add('Opera', 2.3)
pie_chart.render_to_file('bar_chart.svg')
```

执行后会生成文件 bar_chart.svg，效果如图 5-13 所示。

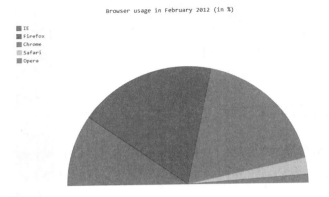

图 5-13　文件 bar_chart.svg 的执行效果

5.1.11 使用库 pandas、numpy 和 matplotlib 绘制饼状图

请看下面的实例文件 pie13.py，功能是使用库 pandas、numpy 和 matplotlib 绘制饼状图。

```
import pandas as pd
import numpy as np
import matplotlib.pyplot as plt

df = pd.DataFrame(3 * np.random.rand(4), index=['a', 'b', 'c', 'd'],
columns=['x'])
df.plot.pie(subplots=True)
plt.show()
```

执行后的效果如图 5-14 所示。

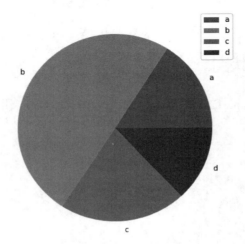

图 5-14　执行后的效果

5.2　爬取热门电影信息并制作可视化分析饼状图

本实例的功能是爬取某电影网的热门电影信息，并将爬取的电影信息保存到 MySQL 数据库中，然后使用 Matplotlib 绘制电影信息的饼状统计图。

5.2.1　创建 MySQL 数据库

扫码观看本节视频讲解

编写文件 myPymysql.py，功能是使用 pymysql 建立和指定 MySQL 数据库的连接，并创建指定选项的数据库表。文件 myPymysql.py 的主要实现代码如下：

```
# 获取 logger 的实例
logger = logging.getLogger("myPymysql")
# 指定 logger 的输出格式
formatter = logging.Formatter('%(asctime)s %(levelname)s %(message)s')
```

```python
# 文件日志,终端日志
file_handler = logging.FileHandler("myPymysql")
file_handler.setFormatter(formatter)

# 设置默认的级别
logger.setLevel(logging.INFO)
logger.addHandler(file_handler)

class DBHelper:
    def __init__(self, host="127.0.0.1", user='root',
                 pwd='66688888',db='testdb',port=3306,
                 charset='utf-8'):
        self.host = host
        self.user = user
        self.port = port
        self.passwd = pwd
        self.db = db
        self.charset = charset
        self.conn = None
        self.cur = None

    def connectDataBase(self):
        """
        连接数据库
        """
        try:
            self.conn =pymysql.connect(host="127.0.0.1",
                user='root',password="66688888",db="testdb",charset="utf8")

        except:
            logger.error("connectDataBase Error")
            return False

        self.cur = self.conn.cursor()
        return True

    def execute(self, sql, params=None):
        """
        执行一般的SQL语句
        """
        if self.connectDataBase() == False:
            return False

        try:
            if self.conn and self.cur:
                self.cur.execute(sql, params)
                self.conn.commit()
        except:
            logger.error("execute"+sql)
            logger.error("params",params)
            return False
        return True

    def fetchCount(self, sql, params=None):
        if self.connectDataBase() == False:
```

```
        return -1
    self.execute(sql, params)
    return self.cur.fetchone()  # 返回操作数据库操作得到一条结果数据

  def myClose(self):
    if self.cur:
      self.cur.close()
    if self.conn:
      self.conn.close()
    return True
if __name__ == '__main__':
  dbhelper = DBHelper()

  sql = "create table maoyan(title varchar(50),actor varchar(200),time varchar(100));"
  result = dbhelper.execute(sql, None)
  if result == True:
    print("创建表成功")
  else:
    print("创建表失败")
  dbhelper.myClose()
  logger.removeHandler(file_handler)
```

执行后会在名为 testdb 对数据库中创建名为 maoyan 的数据库表，如图 5-15 所示。

图 5-15 创建的 MySQL 数据库表

5.2.2 爬取并分析电影数据

编写文件域名.py，功能是抓取指定网页的电影信息，将抓取到的数据添加到 MySQL 数据库中。然后建立和 MySQL 数据库的连接，并使用 matplotlib 将数据库中的电影数据绘制国别类别的统计饼状图。文件域名.py 的主要实现代码如下：

```
import logging

# 获取 logger 的实例
logger = logging.getLogger("maoyan")
# 指定 logger 的输出格式
formatter = logging.Formatter('%(asctime)s %(levelname)s %(message)s')
# 文件日志,终端日志
file_handler = logging.FileHandler("域名.txt")
file_handler.setFormatter(formatter)
```

```python
# 设置默认的级别
logger.setLevel(logging.INFO)
logger.addHandler(file_handler)

def get_one_page(url):
    """
    发起Http请求,获取Response的响应结果
    """
    ua_headers = {"User-Agent":"Mozilla/5.0 (Macintosh; U; Intel Mac OS X 10_6_8; en-us) AppleWebKit/534.50 (KHTML, like Gecko) Version/5.1 Safari/534.50"}
    reponse = requests.get(url,headers=ua_headers)
    if reponse.status_code == 200: #ok
        return reponse.text
    return None

def write_to_sql(item):
    """
    把数据写入数据库
    """
    dbhelper = myPymysql.DBHelper()
    title_data = item['title']
    actor_data = item['actor']
    time_data = item['time']
    sql = "INSERT INTO testdb.maoyan(title,actor,time) VALUES (%s,%s,%s);"
    params = (title_data, actor_data, time_data)
    result = dbhelper.execute(sql, params)
    if result == True:
        print("插入成功")
    else:
        logger.error("execute: "+sql)
        logger.error("params: ",params)
        logger.error("插入失败")
        print("插入失败")

def parse_one_page(html):
    """
    从获取到的html页面中提取真实想要存储的数据:
    电影名,主演,上映时间
    """
    pattern = re.compile('<p class="name">.*?title="([\s\S]*?)"[\s\S]*?<p class="star">([\s\S]*?)</p>[\s\S]*?<p class="releasetime">([\s\S]*?)</p>')
    items = re.findall(pattern,html)

    # yield在返回时会保存当前的函数执行状态
    for item in items:
        yield {
            'title':item[0].strip(),
            'actor':item[1].strip(),
            'time':item[2].strip()
        }

import matplotlib.pyplot as plt

def analysisCounry():
```

```python
    # 从数据库表中查询出每个国家的电影数量并进行分析
    dbhelper = myPymysql.DBHelper()
    # fetchCount
    Total = dbhelper.fetchCount("SELECT count(*) FROM 'testdb'.'maoyan';")
    Am = dbhelper.fetchCount('SELECT count(*) FROM 'testdb'.'maoyan' WHERE time like "%美国%";')
    Ch = dbhelper.fetchCount('SELECT count(*) FROM 'testdb'.'maoyan' WHERE time like "%中国%";')
    Jp = dbhelper.fetchCount('SELECT count(*) FROM 'testdb'.'maoyan' WHERE time like "%日本%";')
    Other = Total[0] - Am[0] - Ch[0] - Jp[0]
    sizes = Am[0], Ch[0], Jp[0], Other
    labels = 'America','China','Japan','Others'
    colors = 'Yellow','Red','Black','Green'
    explode = 0,0,0,0
    # 画出统计图表的饼状图
    plt.pie(sizes,explode=explode,labels=labels,
        colors=colors, autopct="%1.1f%%", shadow=True)
    plt.show()


def CrawlMovieInfo(lock, offset):
    """
    抓取电影的电影名,主演,上映时间
    """
    url = 'http://域名.com/board/4?offset='+str(offset)
    # 抓取当前的页面
    html = get_one_page(url)
    #print(html)

    # 这里的 for
    for item in parse_one_page(html):
        lock.acquire()
        #write_to_file(item)
        write_to_sql(item)
        lock.release()

    # 每次下载完一个页面,随机等待 1~3 秒再去抓取下一个页面
    #time.sleep(random.randint(1,3))


if __name__ == "__main__":
    analysisCounry()
    # 把页面做 10 次的抓取,每个页面都是一个独立的入口
    from multiprocessing import Manager
    # 进程池中不能用这个 lock

    # 进程池之间的 lock 需要用 Manager 中 lock
    manager = Manager()
    lock = manager.Lock()

    # 使用 functools.partial 对函数做一层包装,从而把这把锁传递进进程池
    #这样进程池内就有一把锁可以控制执行流程
    partial_CrawlMovieInfo = functools.partial(CrawlMovieInfo, lock)
    pool = Pool()
```

```
pool.map(partial_CrawlMovieInfo, [i*10 for i in range(10)])
pool.close()
pool.join()
logger.removeHandler(file_handler)
```

执行后会将抓取的电影信息添加到数据库中，会根据数据库数据绘制饼状统计图，如图 5-16 所示。

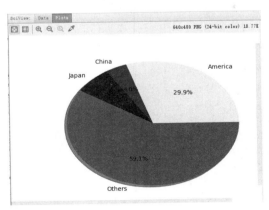

图 5-16　电影统计信息饼状图

5.3　机器学习实战：Scikit-Learn 聚类分析并可视化处理

Scikit-Learn 也可写为 sklearn，是一个开源的基于 Python 语言的机器学习工具包。Scikit-Learn 通过 numpy、scipy 和 matplotlib 等数值计算的库实现高效的算法应用，并且涵盖了几乎所有主流机器学习算法。在本节的内容中，编写实例文件 piechart-kmeans.py，使用 Scikit-Learn 聚类分析饼状图，并统计该饼状图中各种颜色所占的比例，然后可视化展示统计结果。

扫码观看本节视频讲解

5.3.1　准备饼状图

预先准备饼状图素材文件 timg.jpg，效果如图 5-17 所示。

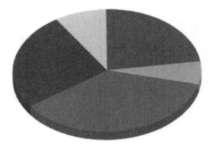

图 5-17　饼状图素材文件 timg.jpg

5.3.2 聚类处理

聚类就是将一组数据按照相识度分割成几类，也就是归类算法。在聚类处理之前，需要先加载饼状图素材文件 timg.jpg，然后将这个图像矩阵转换为像素的列表，并调用聚类算法实现聚类处理。Means(聚类)算法的一个问题在于需要用户事先指定聚类中心的个数。也有一些聚类算法可以自动选择最佳的聚类中心数 K(如 XMeans)，不过不在这里的讨论范围之内。本实例的思路是给定 M×N 个像素点，每个像素点有 R、G、B 三个特征，将每个像素点视为一个数据点进行聚类。

```python
# 加载图像
image_base = cv.imread("timg.jpg")
# 初始化图像像素列表
image = image_base.reshape((image_base.shape[0] * image_base.shape[1], 3))
k = 7  # 聚类的类别个数+1,1为白色底色类别,每次都要重新定义
iterations = 4  # 并发数 4
iteration = 300  # 聚类最大循环次数
clt = KMeans(n_clusters=k, n_jobs=iterations, max_iter=iteration)#kmeans 聚类
clt.fit(image)
hist, label = centroid_histogram(clt)
print(hist, label)
clusters = clt.cluster_centers_
if label>-1:
    clt.cluster_centers_ = delete(clt.cluster_centers_, label, axis=0)
```

- ▶ fit: 将 KMeans 应用到给定数据点的集合。
- ▶ clt.labels：返回每个数据点的标签，即每个数据点经过聚类后，被划分在哪一个 cluster 中，如标号 0、1、2、3 等。
- ▶ clt.cluster_centers：每个聚类中心的平均度量值，这里的度量值可以理解为聚类时用于计算的特征，比如这里的 R、G、B 三个值，每个聚类中心的 R、G、B 三个值的平均被视为该聚类的中心。

5.3.3 生成统计柱状图

(1) 编写函数 centroid_histogram( )生成一个柱状图，可视化展示每个聚类中心的数据点数(所占的比例)。函数 centroid_histogram( )的具体实现代码如下：

```python
# 根据聚类的质心,确定直方图
def centroid_histogram(clt):
    # 抓取不同簇的数量并创建直方图
    # 基于分配给每个群集的像素数
    numLabels = np.arange(0, len(np.unique(clt.labels_)) + 1)
    clustercenters = clt.cluster_centers_
    count = 0
    label = -1
    for i in clustercenters:
        # print(i.tolist())
```

```
        center = i.tolist()
        if center[0] > 200 and center[1] > 200 and center[2] > 200:    #判断白色底
色类簇中心点
            label = count
        count += 1

if label > -1:
    labels = clt.labels_.tolist()
    labelsnew = []
    for i in labels:
        if i != label:
            labelsnew.append(i)         #重新生成一个除去白色底色的列表

    dd = numLabels.tolist()
    dd.remove(label)                    #移除白色底色的聚类标签
    clt.labels_ = np.array(labelsnew)
    numLabels = np.array(dd)
    (hist, _) = np.histogram(clt.labels_, bins=numLabels)
else:
    (hist, _) = np.histogram(clt.labels_, bins=numLabels)
# normalize 直方图,使其总和为1
hist = hist.astype("float")
hist /= hist.sum()
return hist, label
```

(2) 编写函数 plot_colors( ),功能是初始化表示相对频率的每种颜色的柱状图。函数 plot_colors( )的具体实现代码如下：

```
def plot_colors(hist, centroids):
    #
    bar = np.zeros((50, 300, 3), dtype="uint8")
    startX = 0

    # 循环遍历每个类簇的百分比和颜色,每个集群
    for (percent, color) in zip(hist, centroids):
        # 绘制每个类簇的相对百分比
        print(percent, color)
        print(str(percent))
        endX = startX + (percent * 300)
        cv.rectangle(bar, (int(startX), 0), (int(endX), 50),
            color.astype("uint8").tolist(), -1)
        cv.putText(bar, str(round(percent * 100, 1)) + '%', (int(startX), 20),
cv.FONT_HERSHEY_SCRIPT_SIMPLEX, 0.4,
            (0, 0, 0), 1)
        startX = endX

return bar
```

执行后会实现聚类处理,并可视化展示素材饼状图中的每种颜色的百分比。执行效果如图 5-18 所示。

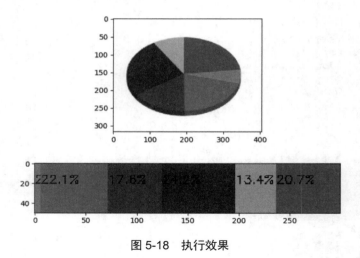

图 5-18　执行效果

5.4　可视化展示名著《西游记》中出现频率最多的文字

在数据可视化分析领域中，文字识别和统计处理是最常用的一种应用情形。在本节的内容中，将在记事本文件中保存四大名著之一的《西游记》电子书，然后对这个记事本文件进行计数测试，最后使用饼状图展示出现频率最多的文字。

扫码观看本节视频讲解

5.4.1　单元测试文件

单元测试是指对软件中的最小可测试单元进行检查和验证。总的来说，单元就是人为规定的最小的被测功能模块。单元测试是在软件开发过程中要进行的最低级别的测试活动，软件的独立单元将在与程序的其他部分相隔离的情况下进行测试。在本项目中的 test 目录下保存了单元测试文件，各个文件的主要功能如下。

- ▶ test_arc.py：实现 matplotlib.pyplot 绘图测试。
- ▶ test_embedded_tk.py：实现 tkinter 窗体内嵌 pyplot 测试。
- ▶ test_thread.py：实现多线程绘图测试。
- ▶ wordCount.py：实现文本文件计数测试。
- ▶ wordGenerate.py：实现生成测试所用文本文件。

5.4.2　GUI 界面

本项目是使用 Python GUI 界面库 tkinter 实现的，为了提高统计小说文字的效率，使用内置库 hreading 实现多线程处理。编写程序文件 MainWindow.py 实现本项目的用户界面，具体实现流程如下。

(1) 编写类 MainWindow 实现主窗体，在顶部设计一个 "打开文件" 菜单。对应的实现代码如下：

```
class MainWindow:
    def __init__(self):
        self.root = tk.Tk()    # 创建主窗体
        self.root.geometry("600x500")
        menubar = tk.Menu(self.root)
        self.canvas = tk.Canvas(self.root)
        menubar.add_command(label='打开文件', command=self.open_file_dialog)
        self.root['menu'] = menubar
        self.expected_percentage = 0.8
```

(2) 编写函数 open_file_dialog(self)，单击"打开文件"菜单后会弹出打开文件对话框。函数 open_file_dialog(self)的具体实现代码如下：

```
def open_file_dialog(self):
    file = tk.filedialog.askopenfilename(title='打开文件', filetypes=[('text',
'*.txt'), ('All Files', '*')])
    self.dialog = InputDialog.InputDialog(self)
    self.dialog.mainloop()
    load_thread = threading.Thread(target=self.create_matplotlib, args=(file,))
    load_thread.start()
```

(3) 编写函数 create_matplotlib(self, f)，功能是使用 matplotlib 库绘制饼状图，首先设置字体为 SimHei，然后使用函数 open( )打开要读取的文件，通过循环统计记事本文件中的所有文字，最终根据统计结果绘制饼状图。为了提高效率，特意使用了以下两种机制。

- 利用 re 模块进行正则匹配筛选文本，简化代码编写量。
- 利用 threading 模块实现多线程读取文件并进行文字统计，从而不影响窗体循环，使得用户可以继续进行操作。

函数 create_matplotlib(self, f)的具体实现代码如下：

```
def create_matplotlib(self, f):
    plt.rcParams['font.sans-serif'] = ['SimHei']
    plt.rcParams['axes.unicode_minus'] = False
    plt.style.use("ggplot")

    file = open(f, 'r', encoding='utf-8')
    word_list = list(file.read())
    word_count = {}
    total_words = 0
    for word in word_list:
        if re.match('\\w', word):
            total_words += 1
            if word in word_count:
                word_count[word] += 1
            else:
                word_count[word] = 1
    max_list = []
    max_count = []
    now_percentage = 0
    if self.expected_percentage >= 1:
        while len(word_count) != 0:
```

```
            max_count.append(0)
            max_list.append("")
            for word, count in word_count.items():
                if max_count[-1] < count:
                    max_count[-1] = count
                    max_list[-1] = word
            # if max_list[-1] in word_count:
            word_count.pop(max_list[-1])
    else:
        while now_percentage < self.expected_percentage:
            max_count.append(0)
            max_list.append("")
            for word, count in word_count.items():
                if max_count[-1] < count:
                    max_count[-1] = count
                    max_list[-1] = word
            # if max_list[-1] in word_count:
            word_count.pop(max_list[-1])
            now_percentage = 0
            for val in max_count:
                now_percentage += val / total_words

    print(max_list, [val / total_words for val in max_count], sep="\n")

    fig = plt.figure(figsize=(32, 32))
    ax1 = fig.add_subplot(111)
    tuple_builder = [0.0 for index in range(len(max_count))]
    tuple_builder[0] = 0.1
    explode = tuple(tuple_builder)
    ax1.pie(max_count, labels=max_list, autopct='%1.1f%%', explode=explode,
textprops={'size': 'larger'})
    ax1.axis('equal')
    ax1.set_title('文字统计')

    self.create_form(fig)
```

（4）编写函数 create_form(self, figure)，使用 Tkinter 的 canvas 绘制 pyplot，然后利用库 matplotlib 提供的 NavigationToolbar2Tk 将 pyplot 的导航栏内嵌到 Tkinter 窗体中。函数 create_form(self, figure)的具体实现代码如下：

```
def create_form(self, figure):
    self.clear()
    figure_canvas = FigureCanvasTkAgg(figure, self.root)
    self.canvas = figure_canvas.get_tk_widget()

    self.toolbar = NavigationToolbar2Tk(figure_canvas, self.root)
    self.toolbar.update()
    self.canvas.pack(side=tk.TOP, fill=tk.BOTH, expand=1)
```

（5）编写函数 clear(self)销毁创建的工具栏，在主函数中启动 GUI 界面。函数 clear(self)的具体实现代码如下：

```
def clear(self):
    self.canvas.destroy()
    if hasattr(self, 'toolbar'):
```

```
        self.toolbar.destroy()
if __name__ == "__main__":
    mainWindow = MainWindow()
    mainWindow.root.mainloop()
```

执行文件 MainWindow.py 后会启动 GUI 主界面，执行效果如图 5-19 所示。单击 GUI 主界面菜单中的"打开文件"，会弹出"打开文件"对话框，如图 5-20 所示。

图 5-19　GUI 主界面　　　　　　　　图 5-20　"打开文件"对话框

5.4.3　设置所需显示的出现频率

编写文件 InputDialog.py，功能是在"打开文件"对话框中选择一个要统计的小说文件后会弹出一个输入对话框，提示用户输入所需显示的出现频率，然后根据用户输入的频率值统计文字。文件 InputDialog.py 的具体实现代码如下：

```
import tkinter as tk

class InputDialog(tk.Tk):

    def __init__(self, main_window):
        tk.Tk.__init__(self)
        self.main_window = main_window
        tk.Label(self, text="输入所需显示的出现频率百分数:").grid(row=0, column=0)
        self.input = tk.Entry(self, bd=5)
        self.input.insert(0, "20%")
        self.input.grid(row=0, column=1)
        tk.Button(self, text="Commit", command=self.__commit).grid(row=1,
column=0, columnspan=2, padx=5, pady=5)

    def __commit(self):
        self.main_window.expected_percentage =
float(self.input.get().strip("%"))/100
        self.quit()
        self.withdraw()
```

```
if __name__ == '__main__':
    window = MainWindow.MainWindow()
    window.root.mainloop()
```

执行后的效果如图 5-21 所示，这说明本项目默认的统计频率值是 20%。单击 Commit 按钮会在"输出"界面打印显示下面的统计结果，并统计当前文件中频率为 20%的文字的统计饼状图，如图 5-22 所示。

```
['道', '不', '一', '了', '那', '我', '是', '来', '他', '个', '行', '你', '的',
'者', '有', '大', '得', '这', '上', '去']
[0.018718106708697443, 0.01499729364466065, 0.013463692285695805,
0.013073909032418393, 0.012735189087430597, 0.012141152600994713,
0.010975207061715115, 0.010103726600238977, 0.0097275602229172127,
0.009656071597566663, 0.00958968929679016, 0.009308841101197264,
0.009182884940749541, 0.008502040830221309, 0.007553965406310744,
0.007249287666684936, 0.006478231711676136, 0.006445891616426045,
0.006292701691557193, 0.006214404618846446]
```

图 5-21 默认值是 20%

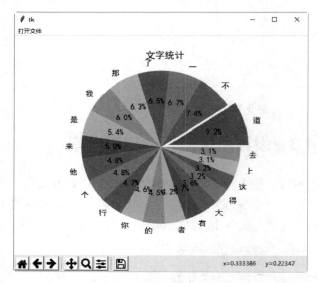

图 5-22 统计饼状图

第 6 章
绘制其他图形

本书前面已经讲解了使用 Python 绘制折线图、散点图、柱状图和饼状图的知识。其实在数据分析可视化应用中,还可以使用其他类型的图形展示数据分析结果。本章将详细讲解使用 Python 绘制其他类型的可视化统计图的知识,为读者步入本书后面知识的学习打下基础。

6.1 绘制雷达图

雷达图也称为网络图、蜘蛛图、星图、蜘蛛网图、不规则多边形、极坐标图或 Kiviat 图。雷达图是以从同一点开始的轴上表示的 3 个或更多个常量或变量的二维图表的形式，这样可以显示多变量数据的图形。在数据可视化分析工作中，经常需要进行数据统计并绘制相关的雷达图。本节将详细讲解在 Python 程序中使用各种库绘制雷达图的知识。

扫码观看本节视频讲解

6.1.1 创建极坐标图

在 Python 程序中，可以使用库 matplotlib 绘制雷达图，在绘制之前需要使用极坐标体系辅助实现。在库 matplotlib 的 pyplot 子库中提供了绘制极坐标图的方法 subplot( )，在通过此方法创建子图时通过设置 projection='polar'便可创建一个极坐标子图,然后调用方法 plot( )在极坐标子图中绘图。请看下面的实例文件 lei01.py，功能是使用库 matplotlib 分别创建一个极坐标图和一个直角坐标子图进行对比。

```
import matplotlib.pyplot as plt
import numpy as np
theta=np.arange(0,2*np.pi,0.02)

ax1 = plt.subplot(121, projection='polar')
ax2 = plt.subplot(122)
ax1.plot(theta,theta/6,'--',lw=2)
ax2.plot(theta,theta/6,'--',lw=2)
plt.show()
```

执行后的效果如图 6-1 所示。

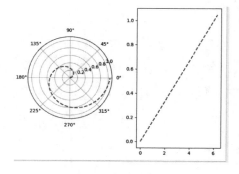

图 6-1 执行后的效果

6.1.2 设置极坐标的正方向

在 Python 程序中使用库 matplotlib 绘制极坐标图时，函数 set_theta_direction( )用于设置

极坐标的正方向。当函数 set_theta_direction( )的参数值为 1、'counterclockwise'或者是'anticlockwise'时,正方向为逆时针方向;当函数 set_theta_direction( )的参数值为-1 或者是'clockwise'时,正方向为顺时针方向。请看下面的实例文件 lei02.py,功能是使用库 matplotlib 创建极坐标图,并且设置了极坐标的正方向。

```python
import matplotlib.pyplot as plt
import numpy as np
theta=np.arange(0,2*np.pi,0.02)
ax1= plt.subplot(121, projection='polar')
ax2= plt.subplot(122, projection='polar')
ax2.set_theta_direction(-1)
ax1.plot(theta,theta/6,'--',lw=2)
ax2.plot(theta,theta/6,'--',lw=2)
plt.show()
```

执行后的效果如图 6-2 所示。

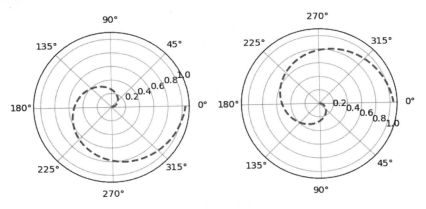

图 6-2　执行后的效果

6.1.3　绘制一个基本的雷达图

请看下面的实例文件 lei03.py,功能是使用库 matplotlib 绘制一个基本的雷达图。

```python
import matplotlib.pyplot as plt
import numpy as np

# 雷达图1 - 极坐标的折线图/填图 - plt.plot()

plt.figure(figsize=(8,4))
ax1= plt.subplot(111, projection='polar')
ax1.set_title('radar map\n')  # 创建标题
ax1.set_rlim(0,12)

data1 = np.random.randint(1,10,10)
data2 = np.random.randint(1,10,10)
data3 = np.random.randint(1,10,10)
theta=np.arange(0, 2*np.pi, 2*np.pi/10)
# 创建数据
```

```
ax1.plot(theta,data1,'.--',label='data1')
ax1.fill(theta,data1,alpha=0.2)
ax1.plot(theta,data2,'.--',label='data2')
ax1.fill(theta,data2,alpha=0.2)
ax1.plot(theta,data3,'.--',label='data3')
ax1.fill(theta,data3,alpha=0.2)
# 绘制雷达线
plt.show()
```

执行后的效果如图 6-3 所示。

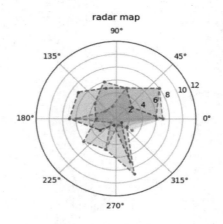

图 6-3 执行后的效果

6.1.4 绘制××战队 2020 绝地求生战绩的雷达图

请看下面的实例文件 lei04.py，功能是使用库 matplotlib 绘制一个雷达图，可视化展示××战队 2020 的绝地求生战绩。

```
import pandas as pd
import numpy as np
import matplotlib.pyplot as plt

plt.rcParams['font.sans-serif'] = ['KaiTi']  # 显示中文
labels = np.array([u'总场次', u'吃鸡数', u'前十数',u'总击杀']) # 标签
dataLenth = 4  # 数据长度
data_radar = np.array([63, 1, 15, 13]) # 数据
angles = np.linspace(0, 2*np.pi, dataLenth, endpoint=False)  # 分割圆周长
data_radar = np.concatenate((data_radar, [data_radar[0]]))  # 闭合
angles = np.concatenate((angles, [angles[0]]))  # 闭合
plt.polar(angles, data_radar, 'bo-', linewidth=1)   # 做极坐标系
plt.thetagrids(angles * 180/np.pi, labels)  # 做标签
plt.fill(angles, data_radar, facecolor='r', alpha=0.25)# 填充
plt.ylim(0, 70)
plt.title(u'XX 战队 2020 的绝地求生战绩')
plt.show()
```

执行后的效果如图 6-4 所示。

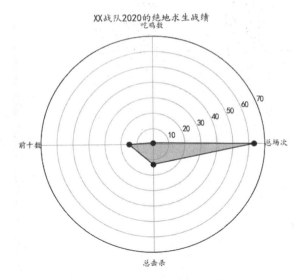

图 6-4　执行后的效果

6.1.5　使用雷达图比较两名研发部同事的能力

请看下面的实例文件 lei05.py，功能是使用库 matplotlib 分别绘制两个雷达图，从 5 个方面(编程能力、沟通技能、专业知识、团队协作、工具掌握)可视化比较两名研发部同事的能力。

```
import numpy as np
from matplotlib import pyplot as plt
fig=plt.figure(figsize=(10,5))
ax1=fig.add_subplot(1,2,1,polar=True)  #设置第一个坐标轴为极坐标体系
ax2=fig.add_subplot(1,2,2,polar=True)  #设置第二个坐标轴为极坐标体系
fig.subplots_adjust(wspace=0.4) #设置子图间的间距,为子图宽度的40%

p1={"编程能力":60,"沟通技能":70,"专业知识":65,"团体协作":75,"工具掌握":80} #创建第一个人的数据
p2={"编程能力":70,"沟通技能":60,"专业知识":75,"团体协作":65,"工具掌握":70} #创建第二个人的数据

data1=np.array([i for i in p1.values()]).astype(int) #提取第一个人的信息
data2=np.array([i for i in p2.values()]).astype(int) #提取第二个人的信息
label=np.array([j for j in p1.keys()]) #提取标签

angle = np.linspace(0, 2*np.pi, len(data1), endpoint=False) #data 里有几个数据就把整圆 360°分成几份
angles = np.concatenate((angle, [angle[0]])) #增加第一个 angle 到所有 angle 里,以实现闭合
data1 = np.concatenate((data1, [data1[0]])) #增加第一个人的第一个 data 到第一个人所有的 data 里,以实现闭合
data2 = np.concatenate((data2, [data2[0]])) #增加第二个人的第一个 data 到第二个人所有的 data 里,以实现闭合
```

```
#设置第一个坐标轴
ax1.set_thetagrids(angles*180/np.pi, label, fontproperties="Microsoft Yahei")
#设置网格标签
ax1.plot(angles,data1,"o-")
ax1.set_theta_zero_location('NW') #设置极坐标0°位置
ax1.set_rlim(0,100) #设置显示的极径范围
ax1.fill(angles,data1,facecolor='g', alpha=0.2) #填充颜色
ax1.set_rlabel_position(255) #设置极径标签位置
ax1.set_title("同事甲",fontproperties="SimHei",fontsize=16) #设置标题

#设置第二个坐标轴
ax2.set_thetagrids(angles*180/np.pi, label, fontproperties="Microsoft Yahei")
#设置网格标签
ax2.plot(angles,data2,"o-")
ax2.set_theta_zero_location('NW') #设置极坐标0°位置
ax2.set_rlim(0,100) #设置显示的极径范围
ax2.fill(angles,data2,facecolor='g', alpha=0.2) #填充颜色
ax2.set_rlabel_position(255) #设置极径标签位置
ax2.set_title("同事乙",fontproperties="SimHei",fontsize=16) #设置标题
plt.show()
```

执行后的效果如图 6-5 所示。

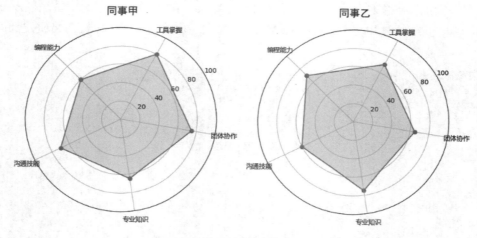

图 6-5 执行后的效果

6.1.6 绘制汽车性能雷达图

请看下面的实例文件 lei06.py，功能是使用库 matplotlib 绘制一个雷达图，可视化展示某款汽车的性能。

```
import numpy as np                                    # 导入科学计算基础包(已安装)
import matplotlib.pyplot as plt                       # 导入绘图库(已安装)
from matplotlib.font_manager import FontProperties    # 导入下载的matplotlib下
可用的中文字体
```

```
# 从文件路径下选择可用的中文字体
font_set=FontProperties(fname="C:\Windows\Fonts\simhei.ttf", size=15)
# 将属性标签放入数组
label=np.array(['耐撞','加速','集气','转向','喷射','漂移'])
#将各属性得分放入数组
data=np.array([3.5,4.5,3.8,5,4,4.5])

# 将2π分为六部分,放入一个数组
angles=np.linspace(0,2*np.pi,len(label),endpoint=False)
data=np.concatenate((data,[data[0]]))
angles=np.concatenate((angles,[angles[0]]))

fig=plt.figure()
ax=fig.add_subplot(111,polar=True)

ax.plot(angles,data,'bo',linewidth=2)
ax.fill(angles,data,facecolor='b',alpha=0.25)
ax.set_thetagrids(angles*180/np.pi,label,fontproperties=font_set)
ax.set_rlim(0,5)
ax.grid(True)

plt.show()
```

执行后的效果如图 6-6 所示。

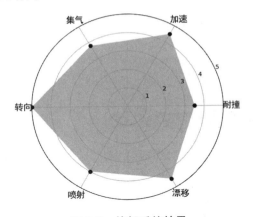

图 6-6　执行后的效果

6.1.7　使用 pygal 绘制雷达图

使用 pygal 绘制雷达图的方法十分简单,只需调用库 pygal 中的 Radar( )方法即可。例如,在下面的实例文件 lei07.py 中,演示了使用方法 Radar( )绘制主流浏览器 V8 基准测试雷达图的过程。

```
import pygal

radar_chart = pygal.Radar()
radar_chart.title = 'V8 基准测试结果'
radar_chart.x_labels = ['Richards', 'DeltaBlue', 'Crypto', 'RayTrace',
'EarleyBoyer', 'RegExp', 'Splay', 'NavierStokes']
```

```
radar_chart.add('Chrome', [6395, 8212, 7520, 7218, 12464, 1660, 2123, 8607])
radar_chart.add('Firefox', [7473, 8099, 11700, 2651, 6361, 1044, 3797, 9450])
radar_chart.add('Opera', [3472, 2933, 4203, 5229, 5810, 1828, 9013, 4669])
radar_chart.add('IE', [43, 41, 59, 79, 144, 136, 34, 102])
radar_chart.render_to_file('bar_chart.svg')
```

执行后会创建生成雷达图文件 bar_chart.svg,打开后的效果如图 6-7 所示。

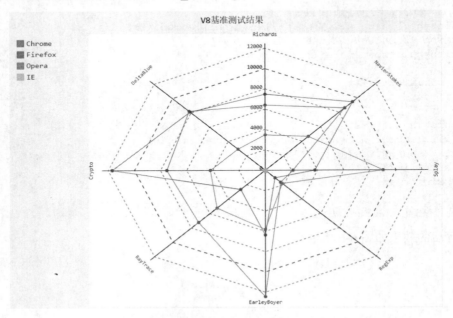

图 6-7　生成的雷达图文件 bar_chart.svg

6.1.8　绘制主流编程语言的雷达图

请看下面的实例文件 lei08.py,功能是使用库 pygal 绘制一个雷达图,从 5 个方面(平台健壮性、语法易用性、社区活跃度、市场份额、未来趋势)可视化比较主流编程语言的特点。

```
import pygal

#准备数据
data =[[5,4.0,5,5,5],
[4.8,2.8,4.8,4.8,4.9],
[4.5,2.9,4.6,4.0,4.9],
[4.0,4.8,4.9,4.0,5],
[3.0,4.2,2.3,3.5,2],
[4.8,4.3,3.9,3.0,4.5]]
#准备标签
labels = ['Java','C','C++','Python','C#','PHP']
#创建 pygal.Radar 对象(雷达图)
radar = pygal.Radar()
#采用循环为雷达图添加数据
for i,per in enumerate(labels):
    radar.add(labels[i],data[i])
```

```
radar.x_labels = ['平台健壮性','语法易用性','社区活跃度','市场份额','未来趋势']
radar.title = '编程语言对比图'
#控制各得分点的大小
radar.dots_size = 8
#设置将图例放在底部
radar.legend_at_bottom = True
#指定将数据图例输出到 SVG 文件中
radar.render_to_file('language.svg')
```

执行后会创建生成雷达图文件 language.svg，打开后的效果如图 6-8 所示。

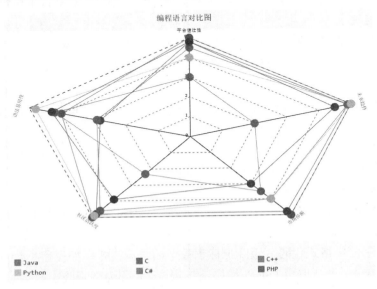

图 6-8　生成的雷达图文件 language.svg

6.2　绘制热力图

热力图是指以特殊高亮的形式显示访客热衷的页面区域和访客所在的地理区域的图示。通过使用热力图，可以显示不可点击区域发生的事情。本节将详细讲解在 Python 程序中使用各种库绘制热力图的知识。

扫码观看本节视频讲解

6.2.1　绘制热力图的函数

在 Python 程序中，通常使用库 Seaborn 中的内置函数 heatmap( )绘制热力图，此函数的语法原型如下：

```
seaborn.heatmap(data, vmin=None, vmax=None,cmap=None, center=None, robust=False, annot=None, fmt='.2g', annot_kws=None,linewidths=0, linecolor='white', cbar=True, cbar_kws=None, cbar_ax=None,square=False, xticklabels='auto', yticklabels='auto', mask=None, ax=None,**kwargs)
```

各类参数的具体说明如下。

(1) 热力图输入数据参数。

data：是一个矩阵数据集，可以是 numpy 的数组(array)，也可以是 pandas 的 DataFrame。如果是 DataFrame，则 df 的 index/column 信息会分别对应到 heatmap 的 columns 和 rows，即 df.index 是热力图的行标，df.columns 是热力图的列标。

(2) 热力图矩阵块颜色参数。

- vmax/vmin：分别是热力图的颜色取值最大和最小范围，默认是根据 data 数据表里的取值确定。
- cmap：从数字到色彩空间的映射，取值是 matplotlib 包里的 colormap 名称或颜色对象，或者表示颜色的列表；该参数默认值根据 center 参数设定。
- center：数据表取值有差异时，设置热力图的色彩中心对齐值；通过设置 center 值，可以调整生成的图像颜色的整体深浅；设置 center 数据时，如果有数据溢出，则手动设置的 vmax、vmin 会自动改变。
- robust：默认取值 False；如果是 False，且没设定 vmin 和 vmax 的值，热力图的颜色映射范围根据具有鲁棒性的分位数设定，而不是用极值设定。

(3) 热力图矩阵块注释参数。

- annot(annotate 的缩写)：默认取值 False；如果是 True 则在热力图每个方格写入数据；如果是矩阵则在热力图中的每个方格写入该矩阵对应位置数据。
- fmt：字符串格式代码，矩阵上标识数字的数据格式，如保留小数点后几位数字。
- annot_kws：默认取值 False，如果是 True，设置热力图矩阵上数字的大小颜色字体，matplotlib 包 text 类下的字体设置。

(4) 热力图矩阵块之间间隔及间隔线参数。

- linewidths：定义热力图里"表示两两特征关系的矩阵小块"之间的间隔大小。
- linecolor：切分热力图上每个矩阵小块的线的颜色，默认值是 white。

(5) 热力图颜色刻度条参数。

- cbar：是否在热力图侧边绘制颜色刻度条，默认值是 True。
- cbar_kws：热力图侧边绘制颜色刻度条时，相关字体设置，默认值是 None。
- cbar_ax：热力图侧边绘制颜色刻度条时，刻度条位置设置，默认值是 None。

(6) square：设置热力图矩阵小块形状，默认值是 False。

(7) xticklabels/yticklabels：xticklabels：控制每列/每行标签名的输出。默认值是 auto。如果是 True，则以 DataFrame 的列名作为标签名。如果是 False，则不添加行标签名。如果是列表则标签名改为列表中给的内容。如果是整数 K 则在图上每隔 K 个标签进行一次标注。如果是 auto 则自动选择标签的标注间距，将标签名不重叠的部分(或全部)输出。

(8) mask：控制某个矩阵块是否显示出来，默认值是 None。如果是布尔型的 DataFrame，则将 DataFrame 里 True 的位置用白色覆盖掉。

(9) ax：设置作图的坐标轴，一般画多个子图时需要修改不同子图的该值。

6.2.2 绘制一个简单的热力图

请看下面的实例文件 re01.py，功能是使用库 seaborn 绘制一个简单的热力图。

```
import numpy as np
import seaborn as sns
import matplotlib.pyplot as plt
sns.set()
np.random.seed(0)
uniform_data = np.random.rand(10, 12)
ax = sns.heatmap(uniform_data)
plt.show()
```

执行后的效果如图 6-9 所示。

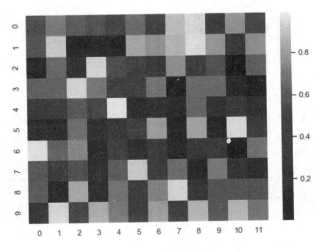

图 6-9　执行效果

6.2.3 使用库 matplotlib 绘制热力图

在 Python 程序中，也可以使用库 matplotlib 绘制热力图。请看下面的实例文件 re02.py，功能是使用库 matplotlib 绘制一个热力图。

```
import numpy as np
import matplotlib.pyplot as plt

points = np.arange(-5,5,0.01)

x,y = np.meshgrid(points,points)
z = np.sqrt(x**2 + y**2)

cmaps = [plt.cm.jet, plt.cm.gray, plt.cm.cool, plt.cm.hot]

fig, axes = plt.subplots(2, 2)

for i, ax in enumerate(axes.ravel()):
```

```
    ax.imshow(z,cmap=cmaps[i])
plt.show()
```

执行后的效果如图 6-10 所示。

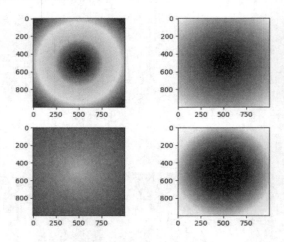

图 6-10 执行后的效果

请看下面的实例文件 re03.py,功能也是使用库 matplotlib 绘制一个热力图。

```
import matplotlib.pyplot as plt
import matplotlib.cm as cm
from matplotlib.colors import LogNorm
import numpy as np
x, y = np.random.rand(10), np.random.rand(10)
z = (np.random.rand(9000000)+np.linspace(0,1, 9000000)).reshape(3000, 3000)
plt.imshow(z+10, extent=(np.amin(x), np.amax(x), np.amin(y), np.amax(y)),
    cmap=cm.hot, norm=LogNorm())
plt.colorbar()
plt.show()
```

执行后的效果如图 6-11 所示。

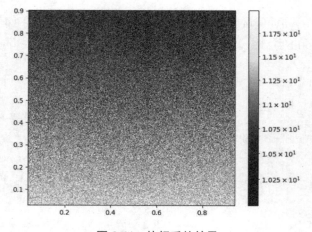

图 6-11 执行后的效果

6.3 将 Excel 文件中的地址信息可视化为交通热力图

请看下面的实例项目，功能是在一个 Excel 文件中保存了一些地址名称，然后将这些文字格式的地址信息转换为坐标信息，并在地图中展示这些地址的实时热力信息。

6.3.1 将地址转换为 JS 格式

扫码观看本节视频讲解

在 Excel 文件 address.xlsx 中保存了文字格式的地址信息，编写 Python 文件 index_address.py 将文件 address.xlsx 中的地址保存到 JS 格式的文件 address.js 中。文件 index_address.py 的具体实现代码如下：

```python
import xlrd
import os
data = xlrd.open_workbook('address.xlsx')
table = data.sheets()[0]
nrows = table.nrows  #行数
ccols = table.col_values(0)
address = []

for rownum in range(1,nrows):
    adr = table.cell(rownum,0).value
    address.append({'name':adr,'address':adr})
print (len(address))

# 保存到文件
address = str(address).encode('utf-8').decode("unicode_escape")
address = address.replace("u", "")
address = '// 自动生成\n' + 'var address = ' + address
with open('./docs/js/data/address.js','a',encoding='utf-8') as jsname:
    jsname.write(address)
```

6.3.2 将 JS 地址转换为坐标

编写文件 amap.js 将 JS 文件 address.js 中的地址转换为坐标格式，具体实现代码如下：

```javascript
var map = new AMap.Map('container',{
    resizeEnable: true,
    zoom: 13
});
map.setCity('北京市');

var geocoder
AMap.plugin(['AMap.ToolBar', 'AMap.Geocoder'],function(){
    geocoder = new AMap.Geocoder({
        city: "010"//城市,默认:"全国"
    });
```

```javascript
        map.addControl(new AMap.ToolBar());
        map.addControl(geocoder);
});
function getMarker (map, model, markers) {
    var address = model.address;
    geocoder.getLocation(address, function(status,result){
        queryNum++
        var marker
        if(status=='complete'&&result.geocodes.length){
            marker = new AMap.Marker({
                map: map,
                position: result.geocodes[0].location
            });
            marker.model = model
            marker.setLabel({//label 默认蓝框白底左上角显示,样式 className 为 amap-marker-label
                offset: new AMap.Pixel(20, 20),//修改 label 相对于 maker 的位置
                content: model.name
            });
            markers.push(marker);

            position.push({
                lng: result.geocodes[0].location.lng,
                lat: result.geocodes[0].location.lat,
                count: position.length + 1,
                ...model
            })
        }else{
            console.log('获取位置失败', address);
        }
        return marker
    })
}
var markers = []
var position = []
var locationCount = 0
var queryNum = 0
function modelsToMap (map, models) {
    markers = []
    locationCount= models ? models.length : 0
    queryNum = 0
    var model
    for (model of models) {
        getMarker(map, model, markers)
    }
}

function load () {
    // address 从 address.js 中获取
    // address 数据格式
    // var address = [{'name': '天安门','address':'xxx 号'},{'name': '水立方','address':'yyy 号'}]
    modelsToMap(map, address)
}
```

```
function downloadPostions() {
    if (queryNum == locationCount) {
        // 调用download.js中的方法
        download('position' + position.length, '// amsp.js生成\n' + 'var postions = ' +JSON.stringify(position))
    }else{
        alert('还没处理完.请稍后...')
    }
}
```

6.3.3 在地图中显示地址的热力信息

申请高德地图 API 的 key，编写文件 index_address.html，在高德地图中可视化展示文件 address.xlsx 中各个地址的热力信息。文件 index_address.html 的具体实现代码如下：

```
<link rel="stylesheet" href="http://cache.amap.com/lbs/static/main1119.css"/>
    <script src="http://webapi.amap.com/maps?v=1.4.0&key=您申请的key值"></script>
    <script type="text/javascript" src="http://cache.amap.com/lbs/static/addToolbar.js"></script>
    <script type="text/javascript" src="./js/data/postions.js"></script>
</head>
<body>
<div id="container"></div>
<div class="button-group">
    <input type="button" class="button" value="显示热力图" onclick="heatmap.show()"/>
    <input type="button" class="button" value="关闭热力图" onclick="heatmap.hide()"/>
</div>
</body>
<script type="text/javascript" src="./js/reli/amap.js"></script>
```

执行后会显示文件 address.xlsx 中各个地址的热力信息，如图 6-12 所示。

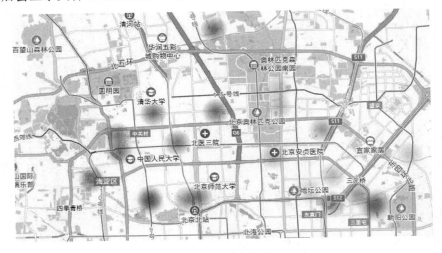

图 6-12 执行效果

6.4 使用热点图可视化展示电视剧的收视率

在本项目实例中,将会抓取互联网电影资料库中的电视剧信息,抓取每部电视剧每一季中各个剧集的收视率,并使用热点图可视化展示每一季每个剧集的收视率。

6.4.1 爬虫爬取电视剧资料

扫码观看本节视频讲解

编写程序文件 film.py 抓取互联网电影资料库中的电视剧信息,首先提示用户输入潮汛电视剧的关键字,然后根据关键字在网站中搜索相关的电视剧,并爬取这个电视剧每一季中每一集的收视率。在文件 film.py 中,实现爬虫抓取功能的代码如下:

```
# 将整个代码放入while循环中,检查用户是否完成了该操作
done = False
while done == False:
    print("可视化显示电视节目收视")

    # 检查URL是否有效,如果有异常则输出显示提示
    validUrl = True
    while validUrl:

        # 用户输入搜索关键字,在IMDB上搜索
        title_search = input("开始在互联网电影资料库搜索电视剧: ")
        title_search = title_search.replace(' ', '+')
        url_search = 'https://www.imdb.com/find?q=' + title_search + '&ref_=nv_sr_sm'

        html_search = requests.get(url_search).text
        soup_search = BeautifulSoup(html_search, 'html.parser')

        try:
            href_list = []
            for link in soup_search.find_all('a', href=True):
                href_list.append(link['href'])

            title_list = []
            i = 0
            for title in href_list:
                if '/title/t' in href_list[i]:
                    title_list.append(href_list[i])
                i += 1

            season = 1
            url1 = 'https://www.imdb.com' + title_list[0]
            url2 = 'episodes?season='
            url3 = str(season)
            url4 = url1 + url2 + url3
            html = requests.get(url4).text
```

```python
        soup = BeautifulSoup(html, 'html.parser')
        t1 = soup.find('h3', attrs={'itemprop': 'name'})
        titleyear = t1.text.split()
        validUrl = False
        print('在 IMDB 上找到了标题!')
    except:
        print('在 IMDB 上找不到此电视剧标题.\n')

# 在 IMDB 上有两种年份方式,一种是检查并相应地删除最后一个字符
if titleyear[-1] == ')':
    year = titleyear[-2] + titleyear[-1]
    del titleyear[-1]
    del titleyear[-1]
else:
    year = titleyear[-1]
    del titleyear[-1]

# 正在提取标题
title = ""
for i in range(0, len(titleyear)):
    if i == 0:
        title += titleyear[i]
    else:
        title += ' ' + titleyear[i]

# 得到季节和年份
t_seasons = []
for option in soup.find_all('option'):
    stemp = option.text.split()
    t_seasons.append(stemp)

c_seasons = [x for x in t_seasons if x]

# 只提取季
seasons = []
for i in range(0, len(c_seasons)):
    if c_seasons[i][0].isnumeric():
        k = int(c_seasons[i][0])
        if k <= 1000:
            seasons.append(c_seasons[i][0])
    else:
        seasons.append(c_seasons[i][0])

print ('Total Seasons:', len(seasons), '\n')
print("开始搜索所有电视剧的收视率...")

# 浏览每季获取的收视率
csvR = []
epsMax = 0
for season in range(1, len(seasons) + 1):
    url4 = url1 + url2 + str(season)
    html = requests.get(url4).text
    soup = BeautifulSoup(html, 'html.parser')

    rating = []
```

```python
        eps = []
        counter = 0
        for div in soup.body.find_all('div', attrs={'class': 'ipl-rating-star small'}):
            rtemp = div.find('span', class_='ipl-rating-star__rating')
            rating.append(rtemp.text)
            counter += 1
            eps.append(counter)

            if epsMax < counter:
                epsMax = counter

            # 收视率
        csvR.append(rating)

        print ("季:", season, 'of', len(seasons))
        print ("剧集:", eps)
        print ("收视率:", rating, "\n")

#收视率列表
tempE = []
i = 1
while i <= epsMax:
    tempE.append(i)
    i += 1

csvE = []
csvE.append(tempE)

print(csvR)
print(csvE)
print('循环完成，创建热图...')
```

6.4.2 使用热点图实现可视化

接下来使用程序文件 film.py 将抓取的电视剧信息保存到 CSV 格式文件 out.csv 中，然后读取 CSV 文件中的收视率数据，并将读取的收视率数据作为参数绘制热点图，将绘制的热点图保存为 PNG 格式的图片文件。在文件 film.py 中，实现热点图可视化处理功能的代码如下：

```python
#创建包含收视率和分级的 CSV 文件
csv_path = r"out.csv"
with open(csv_path, "w", newline="") as f:
    writer = csv.writer(f)
    writer.writerows(csvE)
    writer.writerows(csvR)

# 读取 CSV 文件并将其放入数据帧中
data = pd.read_csv(csv_path)
df = pd.DataFrame(data)
df.index = np.arange(1, len(df) + 1)
```

```python
# 创建一个热图,调整大小,标签,显示
ax = sns.heatmap(df, linewidths=.5, annot=True, cmap="RdYlGn", square=True,
            cbar_kws={'label': '\n\nApplication & Heatmap\nby infosechris@gmail.com'})

plt.title(title + ' ' + year, pad=20, size=16)
plt.ylabel('Seasons', labelpad=30, rotation=0)
plt.xlabel('Episodes', labelpad=10)

ax.xaxis.tick_top()
ax.invert_yaxis()
plt.yticks(rotation=0)
plt.xticks(rotation=0)

fig = plt.gcf()
figsize = fig.get_size_inches()
fig.set_size_inches(figsize * 1.5)

if epsMax >= 100:
    fig.set_figwidth(80)
elif epsMax >= 35:
    fig.set_figwidth(20)
elif epsMax >= 25:
    fig.set_figwidth(20)
elif epsMax >= 16:
    fig.set_figwidth(15)

if season >= 100:
    fig.set_figheight(80)
elif season >= 35:
    fig.set_figheight(20)
elif season >= 25:
    fig.set_figheight(20)
elif season >= 20:
    fig.set_figheight(15)

title = title.replace(':', '')
title = title.replace('?', '')
filename = title + ' ' + year + '.png'

# 如果效果图已经存在则将其删除
ss_path = r"aa\\"
if os.path.exists(ss_path + filename):
    os.remove(ss_path + filename)

# 保存为.PNG 格式图片
fig.savefig(ss_path + filename)

print('Heatmap Image Saved!\n')

# 创建图片完成后打开这个图片
os.startfile(ss_path + filename)

# 询问用户是否已完成应用程序,中断/退出整个 while 循环的唯一方法是在开始时初始化
loopo = True
```

```
            notaskingagain = 0
            while loopo:
                finish = input("继续搜索吗? (Y/N): ")
                if finish == 'N' or finish == 'n':
                    print("程序已经关闭! \n")
                    done = True
                    loopo = False
                elif finish == 'y' or finish == 'y':
                    loopo = False
                    print('\n')
                else:
                    print("没有满足条件的结果，输入 Y 或者 N\n")
                    notaskingagain += 1

                if notaskingagain == 3:
                    print("程序已经关闭! \n")
                    done = True
                    loopo = False
```

执行后会根据输入搜索的关键字打印输出抓取的电视剧信息，并将绘制的热点图保存为一幅 .png 格式的图片文件。例如，输入的关键字是 a，则会打印输出以下爬虫信息，并创建一幅名为 Entourage (2004–2011).png 的热点图文件，效果如图 6-13 所示。

```
可视化显示电视节目收视
开始在互联网电影资料库搜索电视剧： a
在 IMDB 上找到了标题!
Total Seasons: 8

开始搜索所有电视剧的收视率...
季: 1 of 8
剧集: [1, 2, 3, 4, 5, 6, 7, 8]
收视率: ['7.7', '7.7', '8.2', '7.9', '8.2', '8.2', '8.2', '8.4']

季: 2 of 8
剧集: [1, 2, 3, 4, 5, 6, 7, 8, 9, 10, 11, 12, 13, 14]
收视率: ['8.2', '8.7', '8.3', '8.5', '8.3', '8.4', '9.0', '8.1', '8.5', '8.6',
'8.1', '8.4', '9.2', '8.8']

季: 3 of 8
剧集: [1, 2, 3, 4, 5, 6, 7, 8, 9, 10, 11, 12, 13, 14, 15, 16, 17, 18, 19, 20]
收视率: ['8.6', '9.2', '7.8', '8.1', '8.6', '8.8', '8.3', '8.7', '9.1', '8.5',
'8.5', '8.7', '8.3', '8.3', '8.6', '8.5', '8.5', '8.9', '8.4', '8.7']

季: 4 of 8
剧集: [1, 2, 3, 4, 5, 6, 7, 8, 9, 10, 11, 12]
收视率: ['9.0', '8.2', '8.2', '8.7', '8.4', '8.2', '8.6', '8.1', '8.3', '8.2',
'8.1', '8.8']

季: 5 of 8
剧集: [1, 2, 3, 4, 5, 6, 7, 8, 9, 10, 11, 12]
收视率: ['8.2', '8.0', '8.9', '8.4', '8.5', '8.1', '8.9', '8.7', '8.1', '8.1',
'8.9', '9.2']
```

```
季: 6 of 8
剧集: [1, 2, 3, 4, 5, 6, 7, 8, 9, 10, 11, 12]
收视率: ['8.5', '8.3', '8.0', '7.8', '8.1', '8.0', '8.1', '8.3', '8.0', '8.1',
'8.3', '9.3']

季: 7 of 8
剧集: [1, 2, 3, 4, 5, 6, 7, 8, 9, 10]
收视率: ['8.0', '7.9', '7.6', '8.0', '8.2', '8.0', '7.9', '7.9', '8.0', '9.0']

季: 8 of 8
剧集: [1, 2, 3, 4, 5, 6, 7, 8]
收视率: ['8.2', '8.0', '8.1', '7.9', '8.0', '7.8', '8.5', '8.8']

[[['7.7', '7.7', '8.2', '7.9', '8.2', '8.2', '8.2', '8.4'], ['8.2', '8.7', '8.3',
'8.5', '8.3', '8.4', '9.0', '8.1', '8.5', '8.6', '8.1', '8.4', '9.2', '8.8'],
['8.6', '9.2', '7.8', '8.1', '8.6', '8.8', '8.3', '8.7', '9.1', '8.5', '8.5',
'8.7', '8.3', '8.3', '8.6', '8.5', '8.5', '8.9', '8.4', '8.7'], ['9.0', '8.2',
'8.2', '8.7', '8.4', '8.2', '8.6', '8.1', '8.3', '8.2', '8.1', '8.8'], ['8.2',
'8.0', '8.9', '8.4', '8.5', '8.1', '8.9', '8.7', '8.1', '8.1', '8.9', '9.2'],
['8.5', '8.3', '8.0', '7.8', '8.1', '8.0', '8.1', '8.3', '8.0', '8.1', '8.3',
'9.3'], ['8.0', '7.9', '7.6', '8.0', '8.2', '8.0', '7.9', '7.9', '8.0', '9.0'],
['8.2', '8.0', '8.1', '7.9', '8.0', '7.8', '8.5', '8.8']]
[[1, 2, 3, 4, 5, 6, 7, 8, 9, 10, 11, 12, 13, 14, 15, 16, 17, 18, 19, 20]]
循环完成,创建热图...
Heatmap Image Saved!
```

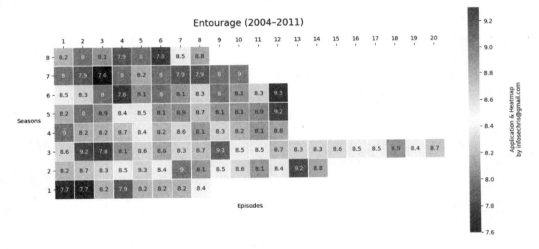

图 6-13 创建的热点图文件 Entourage (2004–2011).png

6.5 行人重识别并绘制行走热力图

　　行人重识别(Person re-identification)也称行人再识别,是利用计算机视觉技术判断图像或者视频序列中是否存在特定行人的技术,被广泛认为是一个图像检索的子问题。给定一个监控行人图像,检

扫码观看本节视频讲解

索跨设备下的该行人图像,旨在弥补固定的摄像头的视觉局限,并可与行人检测/行人跟踪技术相结合,可广泛应用于智能视频监控、智能安保等领域。在本项目实例中,将识别一幅有行人的照片,通过行人重识别后绘制对应的行走热力图。

6.5.1 安装第三方库 pytorch

在 Python 程序中,可以使用第三方库 pytorch 实现行人重识别功能。pytorch 是一个开源的 Python 机器学习库,基于 Torch 实现,主要用于自然语言处理等应用程序。2017 年 1 月,由 Facebook 人工智能研究院(FAIR)基于 Torch 推出了 pytorch,这是一个基于 Python 的可续计算包,主要提供了以下两个功能。

- 具有强大的 GPU 加速的张量计算能力,如 NumPy。
- 包含自动求导系统的深度神经网络。

安装库 pytorch 的过程请读者登录其官方网站查看其官方资料,为节省篇幅,本书不再介绍相关内容。

6.5.2 编写识别程序和绘图程序

编写程序文件 re-identification.py,功能是使用训练模型识别预先准备的素材图片 test.jpg,实现行人重识别后绘制热点图文件 CAM.jpg。程序文件 re-identification.py 的具体实现代码如下:

```python
import numpy as np
import torchvision.models as models
from torchvision import transforms
from torch.autograd import Variable
from torch.nn import functional as F
import cv2
from PIL import Image
net = models.resnet50(pretrained=True)# 你可以用你的训练模型代替这个"resnet18"

net.eval()

features_blobs=[]
def hook_feature(module,input,output):
    features_blobs.append(output.data.cpu().numpy())

net._modules.get('layer4').register_forward_hook(hook_feature)

params = list(net.parameters())
weight_softmax = np.squeeze(params[-2].data.numpy())

def returnCAM(feature_conv,weight_softmax,class_idx):
    size_upsample = (256,256)
    bz,nc,h,w = feature_conv.shape
    output_cam = []
    for idx in class_idx:
```

```python
        cam = weight_softmax[idx].dot(feature_conv.reshape((nc,h*w)))
        cam = cam.reshape(h,w)
        cam = cam-np.min(cam)
        cam_img = cam/np.max(cam)
        cam_img = np.uint8(255*cam_img)
        output_cam.append(cv2.resize(cam_img,size_upsample))
    return output_cam


normalize = transforms.Normalize(
    mean=[0.485, 0.456, 0.406],
    std=[0.229, 0.224, 0.225]
)
preprocess = transforms.Compose([
    transforms.Resize((224,224)),
    transforms.ToTensor(),
    normalize
])

img_pil = Image.open('test.jpg') # use your image of path
img_tensor = preprocess(img_pil)
img_variable = Variable(img_tensor.unsqueeze(0))
logit = net(img_variable)


h_x = F.softmax(logit, dim=1).data.squeeze()
probs, idx = h_x.sort(0, True)
probs = probs.numpy()
idx = idx.numpy()

CAMs = returnCAM(features_blobs[0], weight_softmax, [idx[0]])

img = cv2.imread('test.jpg')
height, width, _ = img.shape
heatmap = cv2.applyColorMap(cv2.resize(CAMs[0],(width, height)), cv2.COLORMAP_JET)
result = heatmap * 0.3 + img * 0.5
cv2.imwrite('CAM.jpg', result)
```

执行后会识别素材图片文件 test.jpg，并将绘制的行人热点图保存在文件 CAM.jpg 中，执行效果如图 6-14 所示。

素材图片文件 test.jpg

绘制的热点图

图 6-14　执行效果

6.6 绘制词云图

"词云"就是通过形成"关键词云层"或"关键词渲染",对网络文本中出现频率较高的"关键词"在视觉上的突出。本节将通过两个实例讲解使用 Python 语言创建词云图的方法。

6.6.1 绘制 B 站词云图

扫码观看本节视频讲解

编写实例文件 BiliBili.py,功能是抓取 B 站排行榜的信息,然后提取排行榜前 50 名视频的所有标签信息,最后将提取的标签文本添加到词云图中。文件 BiliBili.py 的具体实现代码如下:

```
sys.stdout = io.TextIOWrapper( sys.stdout.buffer, encoding='gb18030')#编码
url='https://www.bilibili.com/ranking'#B 站排行榜链接
response=requests.get(url)
html=response.text
video_list=re.findall(r'<a href="(.*?)" target="_blank">', html)#B 站排行榜链接列表
label_list=[]
video_name=re.findall(r'target="_blank" class="title">(.*?)</a><!----><div class="detail"><span class="data-box">',html)#B 站排行榜视频名
video_play=re.findall(r'<i class="b-icon play"></i>(.*?)</span>', html)#播放数
video_view=re.findall(r'<i class="b-icon view"></i>(.*?)</span><a target="_blank"',html)#评论数
video_up=re.findall(r'<i class="b-icon author"></i>(.*?)</span></a>', html)#UP 主
for i in range (0, 100):
    print('%d.'%(i+1), end='')
    print('%-65s'%video_name[i],end='')
    print('up 主: %-15s'%video_up[i], end='')
    print('播放数: %-8s'%video_play[i], end='')
    print('评论数: %s'%video_view[i])#循环输出视频名、UP 主、播放数、评论数
for video in video_list:
    video_response=requests.get(video)
    video_html=video_response.text
    video_label=re.findall(r'target="_blank">(.*?)</a>', video_html)
    for label in video_label:
        label_list.append(label)#把排行榜视频的所有标签添加进 label_list
label_string=" ".join(label_list)#把 label_list 转 string 型
w = wordcloud.WordCloud(width=1000,
                height=700,
                background_color='white',
                font_path='msyh.ttc')
w.generate(label_string)
w.to_file('BiliBili.png')
```

执行后会创建一幅名为 BiliBili.png 的词云图,效果如图 6-15 所示。

图 6-15 词云图效果

6.6.2 绘制知乎词云图

编写实例文件 zhihu.py，功能是抓取知乎热榜的信息，然后提取排行榜前 50 名信息的所有标签信息，最后将提取的标签文本添加到词云图中。文件 zhihu.py 的具体实现代码如下：

```
import matplotlib.pyplot as plt
sys.stdout = io.TextIOWrapper( sys.stdout.buffer, encoding='gb18030')#编码
headers={
    'User-Agent':'Mozilla/5.0 (X11; Linux x86_64) AppleWebKit/537.36 (KHTML, like Gecko) Chrome/66.0.3359.181 Safari/537.36',
    'Referer':'https://www.zhihu.com/',
    'Cookie':'_zap=d39ea996-dcf8-444d-bbb0-2cda0cafc120;
response=requests.get("https://www.zhihu.com/hot",headers=headers)#知乎热榜链接
html=response.text
html_list=re.findall(r'</div></div><div class="HotItem-content"><a href="(.*?)" title="', html)#抓取所有热榜话题链接
print(html_list)
print(len(html_list))
label_list=[]
for i in range(0, 50):
    hot_response=requests.get(html_list[i], headers=headers)#必须要有 headers，否则无法访问
    hot_html=hot_response.text
    hot_label=re.findall(r'keywords" content="(.*?)"/><meta itemProp="answerCount"', hot_html)#抓取所有热词
    hot_name=re.findall(r'><title data-react-helmet="true">(.*?)? - 知乎</title><meta name="viewport"', hot_html)#抓取标题
    print('%d.'%(i+1), end='')
    for i in range(len(hot_label)):
        label_list.append(hot_label[i])
print(label_list)
label_string=" ".join(label_list)#转换为 string 型
print(label_string)
```

```
w = wordcloud.WordCloud(width=1000,
                height=700,
                background_color='white',
                font_path='msyh.ttc')
w.generate(label_string)
w.to_file('zhihu.png')#生成词云图片
```

执行后会创建一幅名为 zhihu.png 的词云图，效果如图 6-16 所示。

图 6-16 词云图效果

6.7 使用热力图可视化展示某城市的房价信息

在本项目实例中，将某城市的某个时间点的房价信息保存在数据库 db.sqlite3 中，然后使用库 django 开发一个 Web 项目，在网页版百度地图中使用热力图可视化展示这些房价信息。

6.7.1 准备数据

扫码观看本节视频讲解

在数据库 db.sqlite3 中保存了某城市的某个时间点的房价信息，有关数据库表的具体设计请参考库 django 文件 models.py，文件 models.py 的具体实现代码如下：

```
from django.db import models
from django.contrib.auth.models import User

class taizhou(models.Model):
    id = models.BigAutoField(primary_key=True)
    name = models.CharField(max_length=100)
    cityid = models.CharField(max_length=10)
    info = models.TextField(blank=True)
    mi2=models.BigIntegerField()
    tel = models.TextField(blank=True)
    avg=models.BigIntegerField(null=True)
```

```
howsell= models.TextField(null=True)
getdate = models.DateTimeField()
GPS_lat = models.TextField(null=True)
GPS_lng = models.TextField(null=True)
```

6.7.2 使用热力图可视化展示信息

编写 Django 项目的模板文件 hot.html,使用百度地图的 API 展示此城市的房价热力图信息,文件 hot.html 的具体实现代码如下:

```
{% load static %}
<!DOCTYPE html>
<html lang="en">
<head>
<!DOCTYPE html>
<html>
<head>
    <meta http-equiv="Content-Type" content="text/html; charset=utf-8" />
    <meta name="viewport" content="initial-scale=1.0, user-scalable=no" />
    <script type="text/javascript"
src="http://api.map.baidu.com/api?v=2.0&ak=Rpu35NRbsPN7gkRWOhT0hTBdYY1BLBp8
"></script>
    <script type="text/javascript"
src="http://api.map.baidu.com/library/Heatmap/2.0/src/Heatmap_min.js"></scr
ipt>
    <title>台州市房产分布热力图</title>
    <style type="text/css">
        ul,li{list-style: none;margin:0;padding:0;float:left;}
        html{height:100%}
        body{height:100%;margin:0px;padding:0px;font-family:"微软雅黑";}
        #container{height:100%;width:100%;}
        #r-result{width:100%;}
    </style>
</head>
<body>
    <div id="container"></div>
    <div id="r-result" style="display:none">
        <input type="button"  onclick="openHeatmap();" value="显示热力图
"/><input type="button"  onclick="closeHeatmap();" value="关闭热力图"/>
    </div>
</body>
</html>
<script type="text/javascript">
    var map = new BMap.Map("container");          // 创建地图实例

    var point = new BMap.Point(121.267705,28.655381);
    map.centerAndZoom(point, 14);                 // 初始化地图,设置中心点坐标和地图级别
    map.setCurrentCity("台州");         //设置当前显示城市
    map.enableScrollWheelZoom(); // 允许滚轮缩放


    var points =[
```

```
{% for taizhou in maplist %}

{"lng":"{{taizhou.GPS_lng}}","lat":"{{taizhou.GPS_lat}}","count":"{{taizhou
.avg}}"},
{% empty %}
 No Data
{% endfor %}

];//这里面添加经纬度


   if(!isSupportCanvas()){
       alert('热力图目前只支持有canvas支持的浏览器,您所使用的浏览器不能使用热力图功能
~')
   }
   //详细的参数,可以查看heatmap.js的文档
https://github.com/pa7/heatmap.js/blob/master/README.md
   //参数说明如下:
   /* visible 热力图是否显示,默认为true
    * opacity 热力的透明度,1-100
    * radius 势力图的每个点的半径大小
    * gradient  {JSON} 热力图的渐变区间 . gradient 如下
    * {
         .2:'rgb(0, 255, 255)',
         .5:'rgb(0, 110, 255)',
         .8:'rgb(100, 0, 255)'
       }
       其中 key 表示插值的位置, 0~1.
         value 为颜色值.
    */
   heatmapOverlay = new
BMapLib.HeatmapOverlay({"radius":100,"visible":true});
   map.addOverlay(heatmapOverlay);
   heatmapOverlay.setDataSet({data:points,max:20000});

   //closeHeatmap();



   //判断浏览器是否支持canvas
   function isSupportCanvas(){
      var elem = document.createElement('canvas');
      return !!(elem.getContext && elem.getContext('2d'));
   }

   function setGradient(){
      /*格式如下
      {
         0:'rgb(102, 255, 0)',
         .5:'rgb(255, 170, 0)',
         1:'rgb(255, 0, 0)'
      }*/
      var gradient = {};
      var colors = document.querySelectorAll("input[type='color']");
      colors = [].slice.call(colors,0);
```

```
        colors.forEach(function(ele){
            gradient[ele.getAttribute("data-key")] = ele.value;
        });
        heatmapOverlay.setOptions({"gradient":gradient});
    }

    function openHeatmap(){
        heatmapOverlay.show();
    }

    function closeHeatmap(){
        heatmapOverlay.hide();
    }
</script>
</body>
</html>
```

通过以下命令启动 Django Web 项目：

```
python manage.py runserver
```

此时会在命令行界面显示以下信息：

```
System check identified no issues (0 silenced).
May 16, 2020 - 21:15:29
Django version 3.0.5, using settings 'hotmap.settings'
Starting development server at http://127.0.0.1:8000/
Quit the server with CTRL-BREAK.
```

在浏览器中输入 http://127.0.0.1:8000/后会在网页中显示对应的可视化热力图效果，执行后的效果如图 6-17 所示。

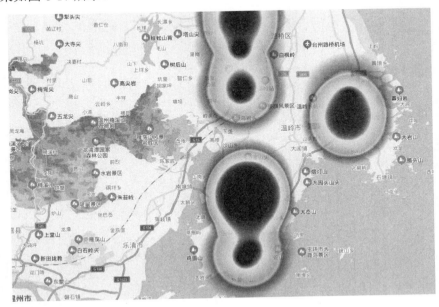

图 6-17　执行后的效果

第 7 章

商业应用：电影票房数据可视化

在当前的市场环境下，去影院看电影仍是消费者休闲娱乐的最主要方式之一，这一点可以从近些年电影市场的高速发展和私人影院的迅速崛起得到佐证。大数据分析电影票房并提取出有关资料，对于电影行业从业者尤为重要。本章将详细讲解提取某专业电影网站电影数据的过程，并根据提取的数据分析电影票房和其他相关资料的过程。

7.1 需求分析

电影这种娱乐载体，出现在众多的消费场景之中，如情侣约会、朋友聚会、闺蜜小聚、公司团建、家庭周末娱乐甚至打发时间。现如今，电影产业的发展更多依靠的是票房和电影院市场的带动。

本项目将抓取××网的电影信息，并用爬虫爬取 2018 年的全年数据和 2019 年的部分(2 月 14 日 22 点之前)的数据，然后进行数据分析。通过使用本系统可以产生以下价值。

扫码观看本节视频讲解

- 电影票房 TOP10：显示本年度总票房前 10 名的电影信息。
- 电影评分 TOP10：显示本年度评分前 10 名的电影信息。
- 电影人气 TOP10：显示本年度点评数量前 10 名的电影信息。
- 每月电影上映数量：显示本年度每月电影的上映数量。
- 每月电影票房：显示本年度每月电影的总票房。
- 各国电影数量 TOP10：显示本年度各国电影数量前 10 名的信息。
- 中外票房对比：显示中外票房对比。
- 名利双收 TOP10：显示本年度名利双收前 10 名的电影信息。
- 叫座不叫好 TOP10：显示本年度叫座不叫好前 10 名的电影信息。
- 电影类型分布：显示本年度所有电影类型的统计信息。

7.2 模块架构

在开发大型应用程序时，模块建构是非常重要的前期准备工作，是关系到整个项目的实现流程是否能顺利完成的关键。本节将根据严格的市场需求分析，得出这个项目的模块结构。本电影信息系统的基本模块架构如图 7-1 所示。

扫码观看本节视频讲解

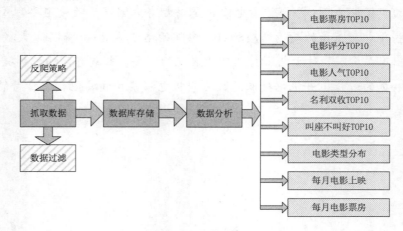

图 7-1 模块架构

7.3 爬虫抓取数据

本节将详细讲解爬取××网电影信息的过程，详细讲解分别爬取 2018 年全年和 2019 年部分(2 月 14 日 22 点)数据的方法。本节只是爬取并分析 2019 年的部分数据，也就是 2019 年 2 月 14 日 22 点之前的数据。

扫码观看本节视频讲解

7.3.1 分析网页

××网的 2018 年电影信息 URL 网页地址是：https://.域名主页 com/films?showType=3&yearId=13&sortId=3&offset=0

××网的 2019 年电影信息 URL 网页地址是：https://.域名主页 com/films?showType=3&yearId=14&sortId=3&offset=0

通过对上述两个分页的分析可以得出以下结果。

- 2018 年电影信息有 184 个分页，每个分页有 30 部电影，但是有评分的只有 10 个分页。
- 2019 年电影信息有 184 个分页，每个分页有 30 部电影，但是有评分的只有 10 个分页。
- 分页参数是 offset，其中第一个分页的值是 30，第 2 个分页的值是 30，第 3 个分页的值是 60，以此类推。
- 2018 年电影信息 URL 地址和 2019 年电影信息 URL 地址的区别是 yearId 编号值，其实 13 表示 2018 年，14 表示 2019 年。

××网某部电影详情页面的 URL 地址是：https://.域名主页 com/films/1200486

在上述地址中，数字 1200486 是这部电影的编号，××网中的每一部电影都有自己的编号，上面的数字 1200486 是电影《我不是药神》的编号。按 F12 键，进入浏览器的开发模式，会发现对评分、评分人数和累计票房等数据进行了文字反爬处理，这些数据都显示为"口口口"的形式，不能直接抓取，如图 7-2 所示。

```
▼<div class="movie-stats-container">
  ▼<div class="movie-index">
      <p class="movie-index-title">用户评分</p>
    ▼<div class="movie-index-content score normal-score">
      ▼<span class="index-left info-num ">
          <span class="stonefont">口.口</span>
        </span>
      ▼<div class="index-right">
        ▼<div class="star-wrapper">
            <div class="star-on" style="width:96%;"></div>
          </div>
        ▼<span class="score-num">
            <span class="stonefont">口口口万</span>
            "人评分"
          </span>
        </div>
      </div>
    </div>
  ▼<div class="movie-index">
      <p class="movie-index-title">累计票房</p>
    ▼<div class="movie-index-content box">
        <span class="stonefont">口口.口口</span>
        <span class="unit">亿</span>
      </div>
    </div>
```

图 7-2 关键数据反爬

7.3.2 破解反爬

打开电影详情页面 https://.域名主页 com/films/1200486，右击查看此网页的源码，然后查找关键字 font-face，找到以下代码：

```
@font-face {
  font-family: stonefont;
  src: url('//vfile.meituan.net/colorstone/
793c4d16ee74ce2c792b9d2fe1d0f4fb3168.eot');
  src: url('//vfile.meituan.net/colorstone/
793c4d16ee74ce2c792b9d2fe1d0f4fb3168.eot?#iefix')
format('embedded-opentype'),
       url('//vfile.meituan.net/colorstone/
4a604c119c4aa8f9585e794730a83fbf2088.woff') format('woff');
}
```

在上述代码中，因为在每次刷新网页后，3 个 URL 网址都会发生变化，所以现在还无法直接匹配信息。接下来需要下载.woff 格式的文件，对其进行破解匹配。破解匹配的基本思路如下。

(1) 首先下载一个字体文件保存到本地(如上面代码中的.woff 格式文件)，如命名为 base.woff，然后人工找出每个数字对应的编码。

(2) 当重新访问网页时，同样也可以把新的字体文件下载下来保存到本地，如叫 base1.woff。网页中的一个数字的编码如为 AAAA，如何确定 AAAA 对应的数字。可以先通过编码 AAAA 找到这个字符在 base1.woff 中的对象，并且把它和 base.woff 中的对象逐个对比，直到找到相同的对象，然后获取这个对象在 base.woff 中的编码，再通过编码确认是哪个数字。

例如，将上述代码中的//vfile.meituan.net/colorstone/4a604c119c4aa8f9585e794730a83fbf2088.woff 输入到浏览器地址，浏览器会自动下载文件 4a604c119c4aa8f9585e794730a83fbf2088.woff。将下载的文件重命名为 base.woff，然后打开网址 http://fontstore.baidu.com/static/editor/index.html，通过此网页打开刚刚下载的文件 base.woff，此时会显示这个文件中的字体对应关系，如图 7-3 所示。

图 7-3　字体对应关系

这说明 uniE05B 代表数字 9，uniF09B 代表数字 0，以此类推。

接下来开始具体编码，实现破解反爬功能，具体实现流程如下。

(1) 编写文件 font_change.py，功能是将上面下载的.woff 格式的文件转换为 XML 文件，这样可以获取爬虫时需要用到的 HTML 标签。文件 font_change.py 的具体实现代码如下：

```
from fontTools.ttLib import TTFont
font = TTFont('base.woff')
font.saveXML('.域名主页 xml')
```

执行后会解析文件 base.woff 的内容，并生成 XML 文件.域名主页 xml。文件.域名主页 xml 的主要内容如下：

```
<GlyphOrder>
  <!-- The 'id' attribute is only for humans; it is ignored when parsed. -->
  <GlyphID id="0" name="glyph00000"/>
  <GlyphID id="1" name="x"/>
  <GlyphID id="2" name="uniE3DE"/>
  <GlyphID id="3" name="uniE88E"/>
  <GlyphID id="4" name="uniE63E"/>
  <GlyphID id="5" name="uniE82E"/>
  <GlyphID id="6" name="uniE94D"/>
  <GlyphID id="7" name="uniF786"/>
  <GlyphID id="8" name="uniE5E6"/>
  <GlyphID id="9" name="uniEEC6"/>
  <GlyphID id="10" name="uniF243"/>
  <GlyphID id="11" name="uniE5C7"/>
</GlyphOrder>

<glyf>
  <TTGlyph name="glyph00000"/><!-- contains no outline data -->

  <TTGlyph name="uniE3DE" xMin="0" yMin="-12" xMax="512" yMax="719">
    <contour>
      <pt x="139" y="173" on="1"/>
      <pt x="150" y="113" on="0"/>
      <pt x="210" y="60" on="0"/>
      <pt x="258" y="60" on="1"/>
      <pt x="300" y="60" on="0"/>
      <pt x="359" y="97" on="0"/>
      <pt x="398" y="159" on="0"/>
      <pt x="412" y="212" on="1"/>
      <pt x="418" y="238" on="0"/>
      <pt x="425" y="292" on="0"/>
      <pt x="425" y="319" on="1"/>
      <pt x="425" y="327" on="1"/>
      <pt x="425" y="331" on="0"/>
      <pt x="424" y="337" on="1"/>
      <pt x="399" y="295" on="0"/>
      <pt x="352" y="269" on="1"/>
      <pt x="308" y="243" on="0"/>
      <pt x="253" y="243" on="1"/>
      <pt x="164" y="243" on="0"/>
      <pt x="42" y="371" on="0"/>
      <pt x="42" y="477" on="1"/>
```

```xml
          <pt x="42" y="586" on="0"/>
          <pt x="169" y="719" on="0"/>
          <pt x="267" y="719" on="1"/>
          <pt x="335" y="719" on="0"/>
          <pt x="452" y="644" on="0"/>
          <pt x="512" y="503" on="0"/>
          <pt x="512" y="373" on="1"/>
          <pt x="512" y="235" on="0"/>
          <pt x="453" y="73" on="0"/>
          <pt x="335" y="-12" on="0"/>
          <pt x="256" y="-12" on="1"/>
          <pt x="171" y="-12" on="0"/>
          <pt x="119" y="34" on="1"/>
          <pt x="65" y="81" on="0"/>
          <pt x="55" y="166" on="1"/>
        </contour>
        <contour>
          <pt x="415" y="481" on="1"/>
          <pt x="415" y="557" on="0"/>
          <pt x="333" y="646" on="0"/>
          <pt x="277" y="646" on="1"/>
          <pt x="218" y="646" on="0"/>
          <pt x="132" y="552" on="0"/>
          <pt x="132" y="474" on="1"/>
          <pt x="132" y="404" on="0"/>
          <pt x="173" y="363" on="1"/>
          <pt x="215" y="320" on="0"/>
          <pt x="336" y="320" on="0"/>
          <pt x="415" y="407" on="0"/>
        </contour>
        <instructions/>
      </TTGlyph>
//省略后面的代码片段
    </glyf>
```

对上述代码的具体说明如下。

▶ 在标签<GlyphOrder>中的内容和前面图 7-3 中的字体对应关系是一一对应的。

▶ 在标签<glyf...>中包含着每个字符对象<TTGlyph>，同样第一个和最后一个不是 0～9 的字符，需要删除。

▶ 在标签<TTGlyph>中包含了坐标点的信息，这些点的功能是描绘字体形状。

(2) 编写文件.域名主页 py，通过函数 get_numbers( )对 XX 的文字进行破解，对应的实现代码如下：

```python
def get_numbers(u):
    cmp = re.compile(",\n               url\('(//.*.woff)'\) format\('woff'\)")
    rst = cmp.findall(u)
    ttf = requests.get("http:" + rst[0], stream=True)
    with open(".域名主页woff", "wb") as pdf:
        for chunk in ttf.iter_content(chunk_size=1024):
            if chunk:
                pdf.write(chunk)
    base_font = TTFont('base.woff')
    maoyanFont = TTFont('.域名主页woff')
```

```python
    maoyan_unicode_list = maoyanFont['cmap'].tables[0].ttFont.getGlyphOrder()
    maoyan_num_list = []
    base_num_list = ['.', '9', '0', '8', '2', '4', '5', '7', '3', '6', '1']
    base_unicode_list = ['x', 'uniE05B', 'uniF09B', 'uniF668', 'uniED4A',
 'uniF140', 'uniE1B2', 'uniF48F', 'uniEB2A', 'uniED40', 'uniF50C']
    for i in range(1, 12):
        maoyan_glyph = maoyanFont['glyf'][maoyan_unicode_list[i]]
        for j in range(11):
            base_glyph = base_font['glyf'][base_unicode_list[j]]
            if maoyan_glyph == base_glyph:
                maoyan_num_list.append(base_num_list[j])
                break
    maoyan_unicode_list[1] = 'uni0078'
    utf8List = [eval(r"'\u" + uni[3:] + "'").encode("utf-8") for uni in
maoyan_unicode_list[1:]]
    utf8last = []
    for i in range(len(utf8List)):
        utf8List[i] = str(utf8List[i], encoding='utf-8')
        utf8last.append(utf8List[i])
    return (maoyan_num_list, utf8last)
```

7.3.3 构造请求头

编写文件.域名主页 py，通过函数 str_to_dict( )构造爬虫所需要的请求头，目的是获取浏览器的访问权限，具体实现代码如下：

```
head = """
Accept:text/html,application/xhtml+xml,application/xml;q=0.9,image/webp,image/apng,*/*;q=0.8
Accept-Encoding:gzip, deflate, br
Accept-Language:zh-CN,zh;q=0.8
Cache-Control:max-age=0
Connection:keep-alive
Host:.域名主页 com
Upgrade-Insecure-Requests:1
Content-Type:application/x-www-form-urlencoded; charset=UTF-8
User-Agent:Mozilla/5.0 (Windows NT 10.0; WOW64) AppleWebKit/537.36 (KHTML, like Gecko) Chrome/59.0.3071.86 Safari/537.36
"""

def str_to_dict(header):
    """
    构造请求头,可以在不同函数里构造不同的请求头
    """
    header_dict = {}
    header = header.split('\n')
    for h in header:
        h = h.strip()
        if h:
            k, v = h.split(':', 1)
            header_dict[k] = v.strip()
    return header_dict
```

7.3.4 实现具体爬虫功能

编写文件.域名主页 py，首先通过函数 get_url( )爬虫获取电影详情页链接，然后通过函数 get_message(url) 爬虫获取电影详情页里的信息。具体实现代码如下：

```
def get_url():
    for i in range(0, 300, 30):
        time.sleep(10)
        url = 'http://.域名主页 com/films?showType=3&yearId=13&sortId=3&offset=' + str(i)
        host = """http://.域名主页 com/films?showType=3&yearId=13&sortId=3&offset=' + str(i)
        """
        header = head + host
        headers = str_to_dict(header)
        response = requests.get(url=url, headers=headers)
        html = response.text
        soup = BeautifulSoup(html, 'html.parser')
        data_1 = soup.find_all('div', {'class': 'channel-detail movie-item-title'})
        data_2 = soup.find_all('div', {'class': 'channel-detail channel-detail-orange'})
        num = 0
        for item in data_1:
            num += 1
            time.sleep(10)
            url_1 = item.select('a')[0]['href']
            if data_2[num-1].get_text() != '暂无评分':
                url = 'http://.域名主页 com' + url_1
                for message in get_message(url):
                    print(message)
                    to_mysql(message)
                print(url)
                print('---------------^^^Film_Message^^^-----------------')
            else:
                print('The Work Is Done')
                break


def get_message(url):
    """
    """
    time.sleep(10)
    data = {}
    host = """refer: http://.域名主页 com/news
    """
    header = head + host
    headers = str_to_dict(header)
    response = requests.get(url=url, headers=headers)
    u = response.text
    # 破解 XX 文字反爬
    (maoyan_num_list, utf8last) = get_numbers(u)
```

```python
# 获取电影信息
soup = BeautifulSoup(u, "html.parser")
mw = soup.find_all('span', {'class': 'stonefont'})
score = soup.find_all('span', {'class': 'score-num'})
unit = soup.find_all('span', {'class': 'unit'})
ell = soup.find_all('li', {'class': 'ellipsis'})
name = soup.find_all('h3', {'class': 'name'})
# 返回电影信息
data["name"] = name[0].get_text()
data["type"] = ell[0].get_text()
data["country"] = ell[1].get_text().split('/')[0].strip().replace('\n', '')
data["length"] = ell[1].get_text().split('/')[1].strip().replace('\n', '')
data["released"] = ell[2].get_text()[:10]
# 因为会出现没有票房的电影,所以这里需要判断
if unit:
    bom = ['分', score[0].get_text().replace('.', '').replace('万', ''), unit[0].get_text()]
    for i in range(len(mw)):
        moviewish = mw[i].get_text().encode('utf-8')
        moviewish = str(moviewish, encoding='utf-8')
        # 通过比对获取反爬文字信息
        for j in range(len(utf8last)):
            moviewish = moviewish.replace(utf8last[j], maoyan_num_list[j])
        if i == 0:
            data["score"] = moviewish + bom[i]
        elif i == 1:
            if '万' in moviewish:
                data["people"] = int(float(moviewish.replace('万', '')) * 10000)
            else:
                data["people"] = int(float(moviewish))
        else:
            if '万' == bom[i]:
                data["box_office"] = int(float(moviewish) * 10000)
            else:
                data["box_office"] = int(float(moviewish) * 100000000)
else:
    bom = ['分', score[0].get_text().replace('.', '').replace('万', ''), 0]
    for i in range(len(mw)):
        moviewish = mw[i].get_text().encode('utf-8')
        moviewish = str(moviewish, encoding='utf-8')
        for j in range(len(utf8last)):
            moviewish = moviewish.replace(utf8last[j], maoyan_num_list[j])
        if i == 0:
            data["score"] = moviewish + bom[i]
        else:
            if '万' in moviewish:
                data["people"] = int(float(moviewish.replace('万', '')) * 10000)
            else:
                data["people"] = int(float(moviewish))
    data["box_office"] = bom[2]
yield data
```

在上述代码中,抓取的 URL 地址参数 yearId 的值是 13,这表示抓取的是 2018 年年底电影信息。如果设置为 yearId=14,则会抓取 2019 年的电影信息。

7.3.5 将爬取的信息保存到数据库

（1）编写文件 maoyan_mysql_1.py，功能是在本地 MySQL 数据库中创建一个名为 maoyan 的数据库，具体实现代码如下：

```
import pymysql

db = pymysql.connect(host='127.0.0.1', user='root', password='66688888', port=3306)
cursor = db.cursor()
cursor.execute("CREATE DATABASE maoyan DEFAULT CHARACTER SET utf8")
db.close()
```

（2）编写文件 maoyan_mysql_2.py，功能是在 MySQL 数据库 maoyan 创建表 films，用于保存抓取到的 2018 年电影信息，具体实现代码如下：

```
import pymysql

db = pymysql.connect(host='127.0.0.1', user='root', password='66688888', port=3306, db='maoyan')
cursor = db.cursor()
sql = 'CREATE TABLE IF NOT EXISTS films (name VARCHAR(255) NOT NULL, type VARCHAR(255) NOT NULL, country VARCHAR(255) NOT NULL, length VARCHAR(255) NOT NULL, released VARCHAR(255) NOT NULL, score VARCHAR(255) NOT NULL, people INT NOT NULL, box_office BIGINT NOT NULL, PRIMARY KEY (name))'
cursor.execute(sql)
db.close()
```

（3）编写文件 maoyan_mysql_2-1.py，功能是在 MySQL 数据库 maoyan 创建表 films1，用于保存抓取到的 2019 年电影信息，具体实现代码如下：

```
import pymysql

db = pymysql.connect(host='127.0.0.1', user='root', password='66688888', port=3306, db='maoyan')
cursor = db.cursor()
sql = 'CREATE TABLE IF NOT EXISTS films1 (name VARCHAR(255) NOT NULL, type VARCHAR(255) NOT NULL, country VARCHAR(255) NOT NULL, length VARCHAR(255) NOT NULL, released VARCHAR(255) NOT NULL, score VARCHAR(255) NOT NULL, people INT NOT NULL, box_office BIGINT NOT NULL, PRIMARY KEY (name))'
cursor.execute(sql)
db.close()
```

（4）在文件 域名主页.py 中编写函数 to_mysql(data)，功能是将抓取到的电影信息保存到数据库，具体实现代码如下：

```
def to_mysql(data):
    """
    信息写入mysql
    """
    table = 'films'
```

```
    keys = ', '.join(data.keys())
    values = ', '.join(['%s'] * len(data))
    db = pymysql.connect(host='localhost', user='root', password='66688888',
port=3306, db='maoyan',charset="utf8")
    cursor = db.cursor()
    sql = 'INSERT INTO {table}({keys}) VALUES ({values})'.format(table=table,
keys=keys, values=values)
    try:
        if cursor.execute(sql, tuple(data.values())):
            print("Successful")
            db.commit()
    except:
        print('Failed')
        db.rollback()
    db.close()
```

上述代码将抓取到的电影信息保存到数据库表 films 中，当然也可以设置将抓取的数据保存到数据库表 films1 中，这要读者根据自己的需要来设置。例如，2018 年的数据被保存到数据库表 films 中，如图 7-4 所示。

	name	type	country	length	released	score	people	box_office
编辑 复制 删除	21克拉	喜剧,爱情	中国大陆	96分钟	2018-04-20	8.3分	124000	110000000
编辑 复制 删除	24小时：末路重生	犯罪,动作	美国,中国大陆,南非	93分钟	2018-01-26	7.9分	11542	9770000
编辑 复制 删除	S4侠降魔记	喜剧	中国大陆	98分钟	2018-01-26	7.9分	2702	3640000
编辑 复制 删除	一个人的课堂	剧情,家庭	中国大陆	105分钟	2018-01-16	9.1分	1353	1370000
编辑 复制 删除	一出好戏	剧情,喜剧	中国大陆	134分钟	2018-08-10	8.2分	937000	1355000000
编辑 复制 删除	一百年很长吗	纪录片	中国大陆	106分钟	2018-12-01	8.1分	1605	1240000
编辑 复制 删除	一纸婚约	剧情,爱情	中国大陆	91分钟	2018-06-08	6.4分	2233	750000
编辑 复制 删除	七号公馆	喜剧,惊悚	中国大陆	94分钟	2018-05-11	5.1分	1428	620000
编辑 复制 删除	三国杀·幻	爱情,动作,悬疑	中国大陆	94分钟	2018-11-09	7.1分	2894	1050000
编辑 复制 删除	三块广告牌	剧情,犯罪	美国,英国	115分钟	2018-03-02	8.6分	34863	64990000
编辑 复制 删除	两个俏公主	动画,奇幻,家庭	中国大陆	89分钟	2018-11-10	7.6分	5267	2230000

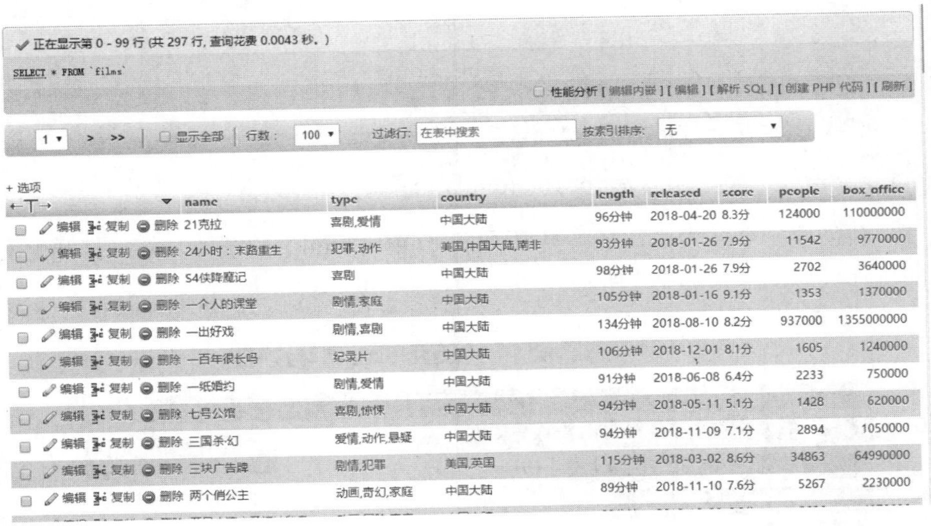

图 7-4　在数据库中保存的电影信息

7.4　数据可视化分析

本节将根据数据库中的电影数据实现大数据分析，逐一分析 2018 年的电影数据和 2019 年的部分(2 月 14 日 22 点前)电影数据。

扫码观看本节视频讲解

7.4.1　电影票房 TOP10

编写文件 movie_box_office_top10.py，功能是统计显示 2019 年

部分(2月14日22点前)总票房前10名的电影信息，具体实现代码如下：

```python
from pyecharts import Bar
import pandas as pd
import numpy as np
import pymysql

conn = pymysql.connect(host='localhost', user='root', password='66688888', port=3306, db='maoyan', charset='utf8')
cursor = conn.cursor()
sql = "select * from films1"
db = pd.read_sql(sql, conn)
df = db.sort_values(by="box_office", ascending=False)
dom = df[['name', 'box_office']]

attr = np.array(dom['name'][0:10])
v1 = np.array(dom['box_office'][0:10])
attr = ["{}".format(i.replace(': 无限战争', '')) for i in attr]
v1 = ["{}".format(float('%.2f' % (float(i) / 10000))) for i in v1]
bar = Bar("2019年电影票房TOP10(万元)(截止到2月14日)", title_pos='center', title_top='18', width=800, height=400)
bar.add("", attr, v1, is_convert=True, xaxis_min=10, yaxis_label_textsize=12, is_yaxis_boundarygap=True, yaxis_interval=0, is_label_show=True, is_legend_show=False, label_pos='right', is_yaxis_inverse=True, is_splitline_show=False)
bar.render("2019年电影票房TOP10.html")
```

执行后会创建一个名为"2019年电影票房TOP10.html"的统计文件，打开此文件会显示对应的统计柱形图，如图7-5所示。

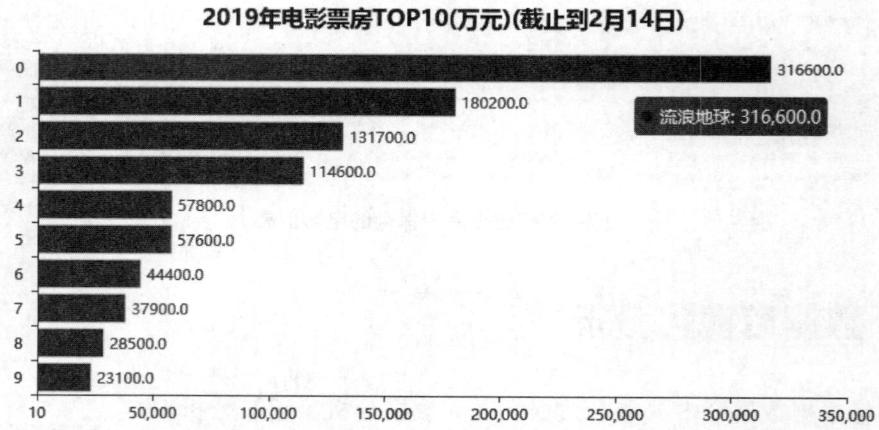

图7-5 电影票房TOP10统计图(2019年部分)

如果在上述代码中设置提取的数据库表是 films，则会统计显示 2018 年的电影票房TOP10数据，如图7-6所示。

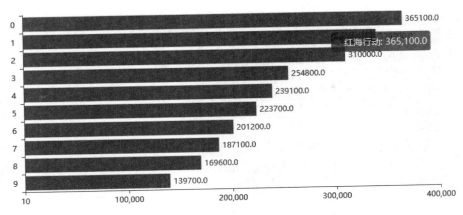

图 7-6 电影票房 TOP10 统计图(2018 年)

7.4.2 电影评分 TOP10

编写文件 movie_score_top10.py，功能是统计分析并显示年度评分前 10 名的电影信息，具体实现代码如下：

```python
from pyecharts import Bar
import pandas as pd
import numpy as np
import pymysql

conn = pymysql.connect(host='localhost', user='root', password='66688888',
port=3306, db='maoyan', charset='utf8')
cursor = conn.cursor()
sql = "select * from films"
db = pd.read_sql(sql, conn)
df = db.sort_values(by="score", ascending=False)
dom = df[['name', 'score']]

v1 = []
for i in dom['score'][0:10]:
    number = float(i.replace('分', ''))
    v1.append(number)
attr = np.array(dom['name'][0:10])
attr = ["{}".format(i.replace(': 致命守护者', '')) for i in attr]

bar = Bar("2019年电影评分TOP10(截止到2月14日)", title_pos='center',
title_top='18', width=800, height=400)
bar.add("", attr, v1, is_convert=True, xaxis_min=8, xaxis_max=9.8,
yaxis_label_textsize=10, is_yaxis_boundarygap=True, yaxis_interval=0,
is_label_show=True, is_legend_show=False, label_pos='right',
is_yaxis_inverse=True, is_splitline_show=False)
bar.render("2019年电影评分TOP10.html")
```

执行后会创建一个名为"2019 年电影评分 TOP10.html"的统计文件，打开此文件会显示对应的统计柱形图。其中 2018 年电影评分 TOP10 统计如图 7-7 所示，2019 年部分电影评分 TOP10 统计如图 7-8 所示。

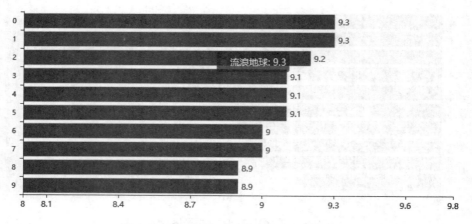

图 7-7　电影评分 TOP10 统计图(2018 年)

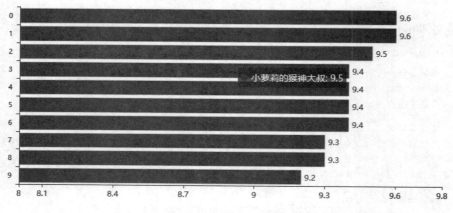

图 7-8　电影评分 TOP10 统计图(2019 年部分)

7.4.3　电影人气 TOP10

编写文件 movie_get_people_top10.py，功能是统计分析并显示年度点评数量前 10 名的电影信息，具体实现代码如下：

```
from pyecharts import Bar
import pandas as pd
import numpy as np
import pymysql

conn = pymysql.connect(host='localhost', user='root', password='66688888',
port=3306, db='maoyan', charset='utf8')
cursor = conn.cursor()
sql = "select * from films1"
db = pd.read_sql(sql, conn)
df = db.sort_values(by="people", ascending=False)
dom = df[['name', 'people']]
```

```
attr = np.array(dom['name'][0:10])
v1 = np.array(dom['people'][0:10])
attr = ["{}".format(i.replace(': 无限战争', '')) for i in attr]
v1 = ["{}".format(float('%.2f' % (float(i) / 1))) for i in v1]
bar = Bar("2019年电影人气TOP10)(截止到2月14日)", title_pos='center',
title_top='18', width=800, height=400)
bar.add("", attr, v1, is_convert=True, xaxis_min=10, yaxis_label_textsize=12,
is_yaxis_boundarygap=True, yaxis_interval=0, is_label_show=True, is_legend_show=
False, label_pos='right', is_yaxis_inverse=True, is_splitline_show=False)
bar.render("2019年电影人气TOP10.html")
```

执行后会创建一个名为"2019年电影人气TOP10.html"的统计文件,打开此文件会显示对应的统计柱形图。其中2018年电影人气TOP10统计如图7-9所示,2019年部分电影票房TOP10统计如图7-10所示。

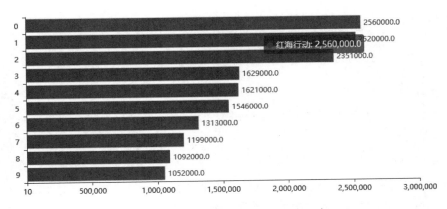

图 7-9　电影人气 TOP10 统计图(2018 年)

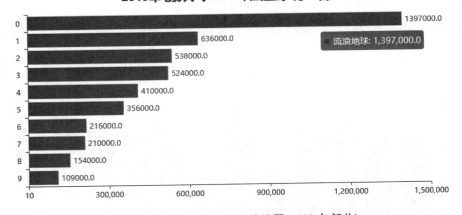

图 7-10　电影人气 TOP10 统计图(2019 年部分)

7.4.4　每月电影上映数量

编写文件 movie_month_update.py,功能是统计分析并显示本年度每月电影的上映数量,具体实现代码如下：

```
conn = pymysql.connect(host='localhost', user='root', password='66688888',
port=3306, db='maoyan', charset='utf8')
cursor = conn.cursor()
sql = "select * from films"
db = pd.read_sql(sql, conn)
df = db.sort_values(by="released", ascending=False)
dom = df[['name', 'released']]
list1 = []
for i in dom['released']:
    place = i.split('-')[1]
    list1.append(place)
db['month'] = list1

month_message = db.groupby(['month'])
month_com = month_message['month'].agg(['count'])
month_com.reset_index(inplace=True)
month_com_last = month_com.sort_index()

attr = ["{}".format(str(i) + '月') for i in range(1, 13)]
v1 = np.array(month_com_last['count'])
v1 = ["{}".format(i) for i in v1]

bar = Bar("2019年每月上映电影数量(截止到2月14日)", title_pos='center',
title_top='18', width=800, height=400)
bar.add("", attr, v1, is_stack=True, yaxis_max=40, is_label_show=True)
bar.render("2019年每月上映电影数量.html")
```

执行后会创建一个名为"2019年每月上映电影数量.html"的统计文件，打开此文件会显示对应的统计柱形图。其中2018年每月上映电影数量统计如图7-11所示，2019年部分每月上映电影数量统计如图7-12所示。因为2019年只统计了两个月的数据，所以上面的代码只能遍历两个月的数据，而不是12个月的，所以需要将遍历行代码改为：

```
attr = ["{}".format(str(i) + '月') for i in range(1, 3)]
```

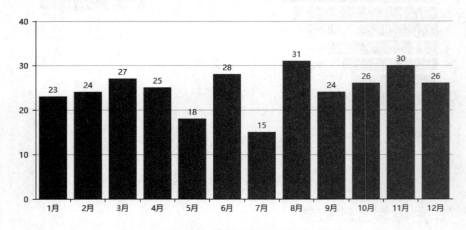

图 7-11　每月上映电影数量统计图(2018年)

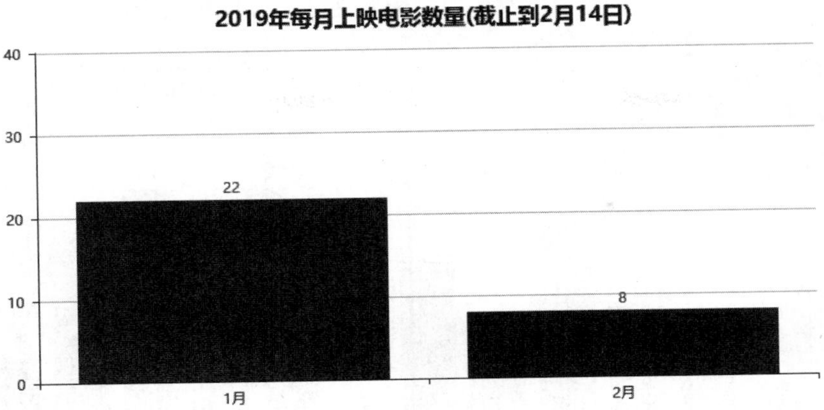

2019年每月上映电影数量(截止到2月14日)

图 7-12　每月上映电影数量统计图(2019 年部分)

7.4.5　每月电影票房

编写文件 movie_month_box_office.py，其功能是统计分析并显示年度每月电影票房的电影信息，具体实现代码如下：

```
conn = pymysql.connect(host='localhost', user='root', password='66688888',
port=3306, db='maoyan', charset='utf8')
cursor = conn.cursor()
sql = "select * from films1"
db = pd.read_sql(sql, conn)
df = db.sort_values(by="released", ascending=False)
dom = df[['name', 'released']]
list1 = []
for i in dom['released']:
    time = i.split('-')[1]
    list1.append(time)
db['month'] = list1

month_message = db.groupby(['month'])
month_com = month_message['box_office'].agg(['sum'])
month_com.reset_index(inplace=True)
month_com_last = month_com.sort_index()

attr = ["{}".format(str(i) + '月') for i in range(1, 3)]
v1 = np.array(month_com_last['sum'])

v1 = ["{}".format(float('%.2f' % (float(i) / 100000000))) for i in v1]
bar = Bar("2019年每月电影票房(亿元)(截止到2月14日)", title_pos='center',
title_top='18', width=800, height=400)
bar.add("", attr, v1, is_stack=True, is_label_show=True)
bar.render("2019年每月电影票房(亿元).html")
```

执行后会创建一个名为 "2019 年每月电影票房(亿元).html" 的统计文件，打开此文件会显示对应的统计柱形图。其中 2018 年每月电影票房统计如图 7-13 所示，2019 年部分每月电影票房统计如图 7-14 所示。因为 2019 年只统计了两个月的数据，所以上面的代码只能

233

遍历两个月的数据，而不是 12 个月的，所以需要将遍历行代码改为：

```
attr = ["{}".format(str(i) + '月') for i in range(1, 3)]
```

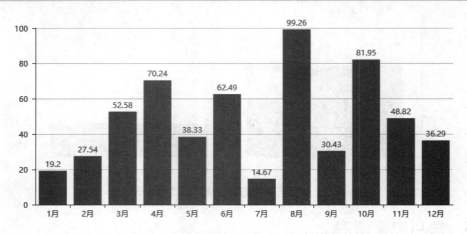

图 7-13　每月电影票房统计图(2018 年)

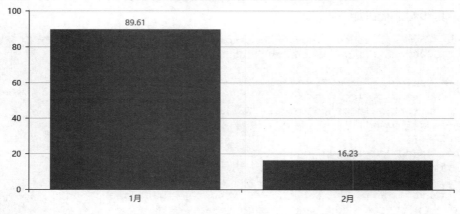

图 7-14　每月电影票房统计图(2019 年部分)

7.4.6　中外票房对比

编写文件 movie_country_box_office.py，功能是统计分析并显示年度中外票房对比的信息，具体实现代码如下：

```
conn = pymysql.connect(host='localhost', user='root', password='66688888',
port=3306, db='maoyan', charset='utf8')
cursor = conn.cursor()
sql = "select * from films"
db = pd.read_sql(sql, conn)
list1 = []
for i in db['country']:
    type1 = i.split(',')[0]
    if type1 in ['中国大陆', '中国香港']:
```

```
        type1 = '中国'
    else:
        type1 = '外国'
    list1.append(type1)
db['country_type'] = list1
country_type_message = db.groupby(['country_type'])
country_type_com = country_type_message['box_office'].agg(['sum'])
country_type_com.reset_index(inplace=True)
country_type_com_last = country_type_com.sort_index()

attr = country_type_com_last['country_type']
v1 = np.array(country_type_com_last['sum'])
v1 = ["{}".format(float('%.2f' % (float(i) / 100000000))) for i in v1]
pie = Pie("2019年中外电影票房对比(亿元)(截止到2月14日)", title_pos='center')
pie.add("", attr, v1, radius=[40, 75], label_pos='right', label_text_color=None,
is_label_show=True, legend_orient="vertical",
legend_pos="left",label_formatter='{c}')
pie.render("2019年中外电影票房对比(亿元).html")
```

执行后会创建一个名为"2019年中外电影票房对比(亿元).html"的统计文件，打开此文件会显示对应的统计柱形图。其中，2018年中外票房对比统计如图7-15所示，2019年部分中外票房对比如图7-16所示。

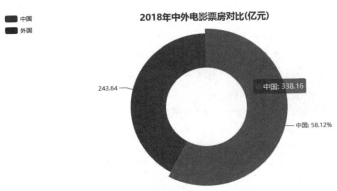

图7-15　中外票房对比统计图(2018年)

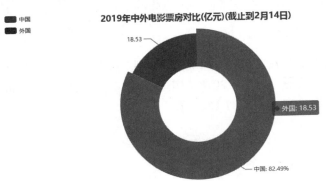

图7-16　中外票房对比统计图(2019年部分)

7.4.7 名利双收 TOP10

编写文件 movie_get_double_top10.py，功能是统计分析并显示年度名利双收 TOP10 的电影信息。名利双收的计算公式是

(某部电影的评分在所有电影评分中的排名+某部电影的票房在所有票房中的排名) /电影总数

文件 movie_get_double_top10.py 的具体实现代码如下：

```python
def my_sum(a, b, c):
    rate = (a + b) / c
    result = float('%.4f' % rate)
    return result


conn = pymysql.connect(host='localhost', user='root', password='66688888',
port=3306, db='maoyan', charset='utf8')
cursor = conn.cursor()
sql = "select * from films1"
db = pd.read_sql(sql, conn)
db['sort_num_money'] = db['box_office'].rank(ascending=0, method='dense')
db['sort_num_score'] = db['score'].rank(ascending=0, method='dense')
db['value'] = db.apply(lambda row: my_sum(row['sort_num_money'],
row['sort_num_score'], len(db.index)), axis=1)
df = db.sort_values(by="value", ascending=True)[0:10]

v1 = ["{}".format('%.2f' % ((1-i) * 100)) for i in df['value']]
attr = np.array(df['name'])
attr = ["{}".format(i.replace(': 无限战争', '').replace(': 全面瓦解', '')) for i
in attr]
bar = Bar("2019 年电影名利双收 TOP10(%) (截止到 2 月 14 日)", title_pos='center',
title_top='18', width=800, height=400)
bar.add("", attr, v1, is_convert=True, xaxis_min=85, xaxis_max=100,
yaxis_label_textsize=12, is_yaxis_boundarygap=True, yaxis_interval=0,
is_label_show=True, is_legend_show=False, label_pos='right',
is_yaxis_inverse=True, is_splitline_show=False)
bar.render("2019 年电影名利双收 TOP10.html")
```

执行后会创建一个名为"2019 年电影名利双收 TOP10.html"的统计文件，打开此文件会显示对应的统计柱形图。其中，2018 年电影名利双收 TOP10 统计如图 7-17 所示，2019 年部分电影名利双收 TOP10 如图 7-18 所示。

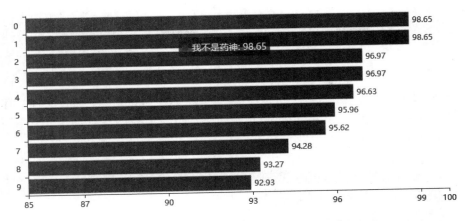

图 7-17 名利双收 TOP10 统计图(2018 年)

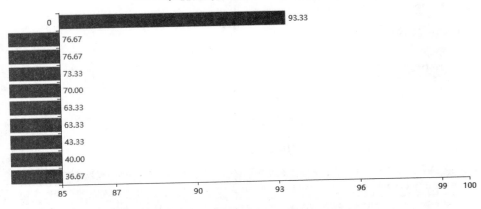

图 7-18 名利双收 TOP10 统计图(2019 年部分)

7.4.8 叫座不叫好 TOP10

编写文件 movie_get_difference_top10.py，功能是统计分析并显示年度叫座不叫好 TOP10 的电影信息。叫座不叫好的计算公式是

(某部电影的票房排名-某部电影的评分排名)/电影总数

文件 movie_get_difference_top10.py 的具体实现代码如下：

```
def my_difference(a, b, c):
    rate = (a - b) / c
    return rate


conn = pymysql.connect(host='localhost', user='root', password='66688888',
port=3306, db='maoyan', charset='utf8')
cursor = conn.cursor()
sql = "select * from films"
a = pd.read_sql(sql, conn)
```

```
a['sort_num_money'] = a['box_office'].rank(ascending=0, method='dense')
a['sort_num_score'] = a['score'].rank(ascending=0, method='dense')
a['value'] = a.apply(lambda row: my_difference(row['sort_num_money'],
row['sort_num_score'], len(a.index)), axis=1)
df = a.sort_values(by="value", ascending=True)[0:9]
```

执行后会创建一个名为"2019年电影叫座不叫好TOP10.html"的统计文件，打开此文件会显示对应的统计柱形图。其中2018年电影叫座不叫好TOP10统计如图7-19所示，2019年部分电影叫座不叫好TOP10如图7-20所示。

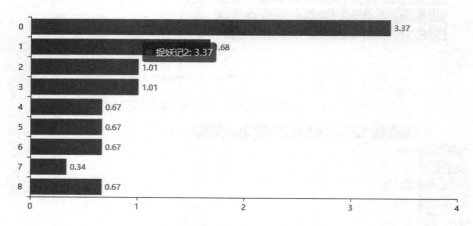

图 7-19　叫座不叫好 TOP10 统计图(2018 年)

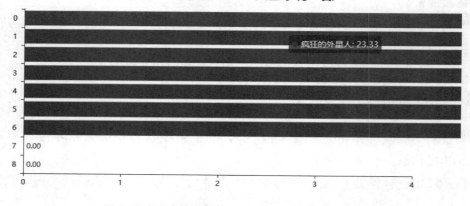

图 7-20　叫座不叫好 TOP10 统计图(2019 年部分)

7.4.9　电影类型分布

编写文件 movie_type.py，功能是统计分析并显示年度电影类型分布的信息，具体实现代码如下：

```
conn = pymysql.connect(host='localhost', user='root', password='66688888',
port=3306, db='maoyan', charset='utf8')
cursor = conn.cursor()
```

```python
sql = "select * from films"
db = pd.read_sql(sql, conn)

dom1 = []
for i in db['type']:
    type1 = i.split(',')
    for j in range(len(type1)):
        if type1[j] in dom1:
            continue
        else:
            dom1.append(type1[j])

dom2 = []
for item in dom1:
    num = 0
    for i in db['type']:
        type2 = i.split(',')
        for j in range(len(type2)):
            if type2[j] == item:
                num += 1
            else:
                continue
    dom2.append(num)

def message():
    for k in range(len(dom2)):
        data = {}
        data['name'] = dom1[k] + ' ' + str(dom2[k])
        data['value'] = dom2[k]
        yield data

data1 = message()
dom3 = []
for item in data1:
    dom3.append(item)

treemap = TreeMap("2019年电影类型分布图(截止到2月14日)", title_pos='center',
title_top='5', width=800, height=400)
treemap.add('2019年电影类型分布', dom3, is_label_show=True, label_pos='inside',
is_legend_show=False)
treemap.render('2019年电影类型分布图.html')
```

执行后会创建一个名为"2019年电影类型分布图.html"的统计文件，打开此文件会显示对应的统计柱形图。其中2018年电影类型分布统计如图7-21所示，2019年部分电影类型分布统计如图7-22所示。

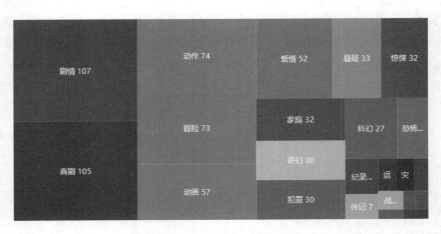

图 7-21　电影类型分布统计图(2018 年)

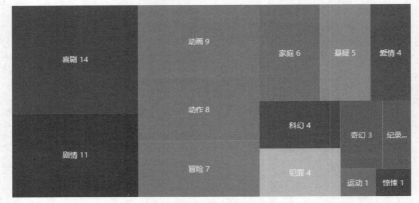

图 7-22　电影类型分布统计图(2019 年部分)

第 8 章

商业应用：房地产市场数据可视化

房价现在已经成为人们最关注的对象之一，些许的风吹草动都会引起大家的注意。本章将详细讲解使用 Python 语言采集主流网站中国内主流城市房价信息的过程，包括新房价格、二手房价格和房租价格，这些采集的数据可以作进一步数据分析。

8.1 背景介绍

随着房价的不断升高，人们对房价的关注度也越来越高，房产投资者希望通过房价数据预判房价走势，从而进行有效的投资，获取收益；因结婚、为小孩上学等需要买房的民众，希望通过房价数据寻找买房的最佳时机，以最适合的价格购买能满足需要的房产。

在当前市场环境下，因为房价水平牵动了大多数人的心，所以各大房产网都上线了"查房价"相关功能模块，以满足购房者/计划购房者经常关注房价行情的需求，从而实现增加产品活跃度、促进购房转化的目的。

整个房产网市场用户群大都一样，主要房源资源和营销方式有所差异。然而，以 X 家和 X 壳为首的房产网巨头公司的房源，由于品牌与质量的优势正快速扩张，市场上的推广费用也越来越高，而购房者迫切希望通过分析找到最精确的房价查询系统。

扫码观看本节视频讲解

8.2 需求分析

本项目将提供国内主流城市、每个区域、每个小区的房价成交情况、关注情况、发展走势，乃至每个小区的解读/评判，以此解决用户购房没有价格依据、无从选择购房时机的问题；满足用户及时了解房价行情，以最合适价格购买最合适位置房产的需求。

通过使用本系统可以产生以下价值。

- ▶ 增加活跃：由于对房价的关注是中长期性质的，不断更新的行情数据可以增加用户活跃度。
- ▶ 促进转化：使用房价数据等帮助用户购房推荐合适的位置与价格，可以提高用户的咨询率与成交率。
- ▶ 减少跳失：若没有此功能，会导致一些购房观望者无从得知房价变化，而最终选择离开。

扫码观看本节视频讲解

8.3 模块架构

查房价系统的基本模块架构如图 8-1 所示。

扫码观看本节视频讲解

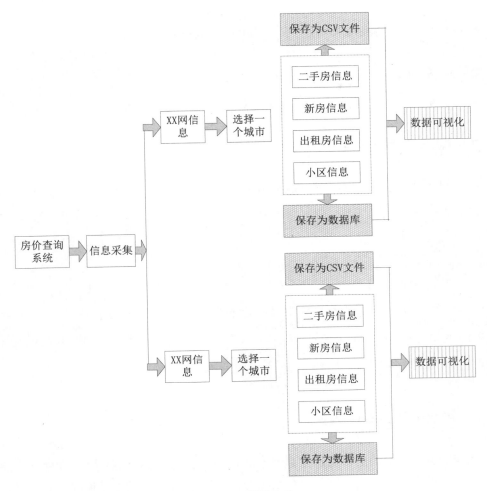

图 8-1　模块架构

8.4　系统设置

在开发一个大型应用程序时，需要模块化开发经常用到的系统设置模块。本节将详细讲解实现本项目系统设置模块的过程。

扫码观看本节视频讲解

8.4.1　选择版本

因为在当前市面中同时存在 Python2 和 Python3 两个版本，所以本系统分别推出了对应的两个实现版本。编写文件 version.py 供用户选择使用不同的 Python 版本，具体实现代码如下：

```
import sys

if sys.version_info < (3, 0):    # 如果小于Python3
```

```
        PYTHON_3 = False
else:
        PYTHON_3 = True

if not PYTHON_3:            # 如果小于Python3
        reload(sys)
        sys.setdefaultencoding("utf-8")
```

8.4.2 保存日志信息

为了便于系统维护，编写文件 log.py 保存使用本系统的日志信息，具体实现代码如下：

```
import logging
from lib.utility.path import LOG_PATH

logger = logging.getLogger(__name__)
logger.setLevel(level=logging.INFO)
handler = logging.FileHandler(LOG_PATH + "/log.txt")
handler.setLevel(logging.INFO)
formatter = logging.Formatter('%(asctime)s - %(levelname)s - %(message)s')
handler.setFormatter(formatter)
logger.addHandler(handler)

if __name__ == '__main__':
        pass
```

8.4.3 设置创建的文件名

本系统能够将抓取的房价信息保存到本地 CSV 文件中，保存 CSV 文件的文件夹的命名机制有日期、城市和房源类型等。编写系统设置文件 path.py，功能是根据不同的机制创建对应的文件夹来保存 CSV 文件。文件 path.py 的具体实现代码如下：

```
def get_root_path():
        file_path = os.path.abspath(inspect.getfile(sys.modules[__name__]))
        parent_path = os.path.dirname(file_path)
        lib_path = os.path.dirname(parent_path)
        root_path = os.path.dirname(lib_path)
        return root_path


def create_data_path():
        root_path = get_root_path()
        data_path = root_path + "/data"
        if not os.path.exists(data_path):
            os.makedirs(data_path)
        return data_path


def create_site_path(site):
        data_path = create_data_path()
        site_path = data_path + "/" + site
```

```python
    if not os.path.exists(site_path):
        os.makedirs(site_path)
    return site_path


def create_city_path(site, city):
    site_path = create_site_path(site)
    city_path = site_path + "/" + city
    if not os.path.exists(city_path):
        os.makedirs(city_path)
    return city_path


def create_date_path(site, city, date):
    city_path = create_city_path(site, city)
    date_path = city_path + "/" + date
    if not os.path.exists(date_path):
        os.makedirs(date_path)
    return date_path


# const for path
ROOT_PATH = get_root_path()
DATA_PATH = ROOT_PATH + "/data"
SAMPLE_PATH = ROOT_PATH + "/sample"
LOG_PATH = ROOT_PATH + "/log"

if __name__ == "__main__":
    create_date_path("lianjia", "sh", "20160912")
    create_date_path("anjuke", "bj", "20160912")
```

8.4.4 设置抓取城市

本系统能够将抓取国内主流一线、二线城市的房价，编写文件 city.py 设置要抓取的城市，实现城市缩写和城市名的映射。如果想抓取其他已有城市的房价，需要把相关城市信息放入文件 city.py 的字典中。文件 city.py 的具体实现代码如下：

```python
cities = {
    'bj': '北京',
    'cd': '成都',
    'cq': '重庆',
    'cs': '长沙',
    'dg': '东莞',
    'dl': '大连',
    'fs': '佛山',
    'gz': '广州',
    'hz': '杭州',
    'hf': '合肥',
    'jn': '济南',
    'nj': '南京',
    'qd': '青岛',
    'sh': '上海',
```

```python
    'sz': '深圳',
    'su': '苏州',
    'sy': '沈阳',
    'tj': '天津',
    'wh': '武汉',
    'xm': '厦门',
    'yt': '烟台',
}


lianjia_cities = cities
beike_cities = cities


def create_prompt_text():
    """
    根据已有城市中英文对照表拼接选择提示信息
    :return: 拼接好的字串
    """
    city_info = list()
    count = 0
    for en_name, ch_name in cities.items():
        count += 1
        city_info.append(en_name)
        city_info.append(": ")
        city_info.append(ch_name)
        if count % 4 == 0:
            city_info.append("\n")
        else:
            city_info.append(", ")
    return 'Which city do you want to crawl?\n' + ''.join(city_info)


def get_chinese_city(en):
    """
    拼音区县名转中文城市名
    :param en: 拼音
    :return: 中文
    """
    return cities.get(en, None)


def get_city():
    city = None
    # 允许用户通过命令直接指定
    if len(sys.argv) < 2:
        print("Wait for your choice.")
        # 让用户选择抓取哪个城市的二手房小区价格数据
        prompt = create_prompt_text()
        # 判断 Python 版本
        if not PYTHON_3:  # 如果小于Python3
            city = raw_input(prompt)
        else:
            city = input(prompt)
```

```python
    elif len(sys.argv) == 2:
        city = str(sys.argv[1])
        print("City is: {0}".format(city))
    else:
        print("At most accept one parameter.")
        exit(1)

    chinese_city = get_chinese_city(city)
    if chinese_city is not None:
        message = 'OK, start to crawl ' + get_chinese_city(city)
        print(message)
        logger.info(message)
    else:
        print("No such city, please check your input.")
        exit(1)
    return city


if __name__ == '__main__':
    print(get_chinese_city("sh"))
```

8.4.5 处理区县信息

因为每个城市都有不同的行政区,所以编写文件 area.py 处理区县信息,具体实现代码如下:

```python
def get_district_url(city, district):
    """
    拼接指定城市的区县url
    :param city: 城市
    :param district: 区县
    :return:
    """
    return "http://{0}.{1}.com/xiaoqu/{2}".format(city, SPIDER_NAME, district)


def get_areas(city, district):
    """
    通过城市和区县名获得下级板块名
    :param city: 城市
    :param district: 区县
    :return: 区县列表
    """
    page = get_district_url(city, district)
    areas = list()
    try:
        headers = create_headers()
        response = requests.get(page, timeout=10, headers=headers)
        html = response.content
        root = etree.HTML(html)
        links = root.xpath(DISTRICT_AREA_XPATH)

        # 针对a标签的list进行处理
```

```python
        for link in links:
            relative_link = link.attrib['href']
            # 去掉最后的"/"
            relative_link = relative_link[:-1]
            # 获取最后一节
            area = relative_link.split("/")[-1]
            # 去掉区县名,防止重复
            if area != district:
                chinese_area = link.text
                chinese_area_dict[area] = chinese_area
                # print(chinese_area)
                areas.append(area)
        return areas
    except Exception as e:
        print(e)
```

然后编写文件 district.py 获取各个区县的详细信息,具体实现代码如下:

```python
chinese_city_district_dict = dict()      # 城市代码和中文名映射
chinese_area_dict = dict()               # 版块代码和中文名映射
area_dict = dict()


def get_chinese_district(en):
    """
    拼音区县名转中文区县名
    :param en: 英文
    :return: 中文
    """
    return chinese_city_district_dict.get(en, None)


def get_districts(city):
    """
    获取各城市的区县中英文对照信息
    :param city: 城市
    :return: 英文区县名列表
    """
    url = 'https://{0}.{1}.com/xiaoqu/'.format(city, SPIDER_NAME)
    headers = create_headers()
    response = requests.get(url, timeout=10, headers=headers)
    html = response.content
    root = etree.HTML(html)
    elements = root.xpath(CITY_DISTRICT_XPATH)
    en_names = list()
    ch_names = list()
    for element in elements:
        link = element.attrib['href']
        en_names.append(link.split('/')[-2])
        ch_names.append(element.text)

        # 打印区县英文和中文名列表
    for index, name in enumerate(en_names):
        chinese_city_district_dict[name] = ch_names[index]
        # print(name + ' -> ' + ch_names[index])
    return en_names
```

8.5 破解反爬机制

在市面中很多站点都设立了反爬机制，防止站点内的信息被爬取。本节将详细讲解本项目破解反爬机制的过程。

8.5.1 定义爬虫基类

扫码观看本节视频讲解

编写文件 base_spider.py 定义爬虫基类，首先设置随机延迟，防止爬虫被禁止；然后设置要爬取的目标站点，下面代码默认抓取的是果壳；最后获取城市列表来选择将要爬取的目标城市。文件 base_spider.py 的具体实现代码如下：

```python
thread_pool_size = 50

# 防止爬虫被禁止，随机延迟设定
# 如果不想 delay，就设定 False
# 具体时间可以修改 random_delay()，由于多线程，建议数值大于 10
RANDOM_DELAY = False
LIANJIA_SPIDER = "lianjia"
BEIKE_SPIDER = "ke"
# SPIDER_NAME = LIANJIA_SPIDER
SPIDER_NAME = BEIKE_SPIDER


class BaseSpider(object):
    @staticmethod
    def random_delay():
        if RANDOM_DELAY:
            time.sleep(random.randint(0, 16))

    def __init__(self, name):
        self.name = name
        if self.name == LIANJIA_SPIDER:
            self.cities = lianjia_cities
        elif self.name == BEIKE_SPIDER:
            self.cities = beike_cities
        else:
            self.cities = None
        # 准备日期信息，爬到的数据存放到日期相关文件夹下
        self.date_string = get_date_string()
        print('Today date is: %s' % self.date_string)

        self.total_num = 0  # 总的小区个数,用于统计
        print("Target site is {0}.com".format(SPIDER_NAME))
        self.mutex = threading.Lock()  # 创建锁

    def create_prompt_text(self):
        """
        根据已有城市中英文对照表拼接选择提示信息
```

```
        :return: 拼接好的字串
        """
        city_info = list()
        count = 0
        for en_name, ch_name in self.cities.items():
            count += 1
            city_info.append(en_name)
            city_info.append(": ")
            city_info.append(ch_name)
            if count % 4 == 0:
                city_info.append("\n")
            else:
                city_info.append(", ")
        return 'Which city do you want to crawl?\n' + ''.join(city_info)

    def get_chinese_city(self, en):
        """
        拼音区县名转中文城市名
        :param en: 拼音
        :return: 中文
        """
        return self.cities.get(en, None)
```

8.5.2 浏览器用户代理

编写文件 headers.py 实现浏览器用户代理功能，具体实现代码如下：

```
USER_AGENTS = [
    "Mozilla/4.0 (compatible; MSIE 6.0; Windows NT 5.1; SV1; AcooBrowser; .NET CLR 1.1.4322; .NET CLR 2.0.50727)",
    "Mozilla/4.0 (compatible; MSIE 7.0; Windows NT 6.0; Acoo Browser; SLCC1; .NET CLR 2.0.50727; Media Center PC 5.0; .NET CLR 3.0.04506)",
    "Mozilla/4.0 (compatible; MSIE 7.0; AOL 9.5; AOLBuild 4337.35; Windows NT 5.1; .NET CLR 1.1.4322; .NET CLR 2.0.50727)",
    "Mozilla/5.0 (Windows; U; MSIE 9.0; Windows NT 9.0; en-US)",
    "Mozilla/5.0 (compatible; MSIE 9.0; Windows NT 6.1; Win64; x64; Trident/5.0; .NET CLR 3.5.30729; .NET CLR 3.0.30729; .NET CLR 2.0.50727; Media Center PC 6.0)",
    "Mozilla/5.0 (compatible; MSIE 8.0; Windows NT 6.0; Trident/4.0; WOW64; Trident/4.0; SLCC2; .NET CLR 2.0.50727; .NET CLR 3.5.30729; .NET CLR 3.0.30729; .NET CLR 1.0.3705; .NET CLR 1.1.4322)",
    "Mozilla/4.0 (compatible; MSIE 7.0b; Windows NT 5.2; .NET CLR 1.1.4322; .NET CLR 2.0.50727; InfoPath.2; .NET CLR 3.0.04506.30)",
    "Mozilla/5.0 (Windows; U; Windows NT 5.1; zh-CN) AppleWebKit/58.15 (KHTML, like Gecko, Safari/419.3) Arora/0.3 (Change: 287 c9dfb30)",
    "Mozilla/5.0 (X11; U; Linux; en-US) AppleWebKit/527+ (KHTML, like Gecko, Safari/419.3) Arora/0.6",
    "Mozilla/5.0 (Windows; U; Windows NT 5.1; en-US; rv:1.8.1.2pre) Gecko/20070215 K-Ninja/2.1.1",
    "Mozilla/5.0 (Windows; U; Windows NT 5.1; zh-CN; rv:1.9) Gecko/20080705 Firefox/3.0 Kapiko/3.0",
    "Mozilla/5.0 (X11; Linux i686; U;) Gecko/20070322 Kazehakase/0.4.5",
    "Mozilla/5.0 (X11; U; Linux i686; en-US; rv:1.9.0.8) Gecko Fedora/1.9.0.8-1.fc10 Kazehakase/0.5.6",
```

```
    "Mozilla/5.0 (Windows NT 6.1; WOW64) AppleWebKit/535.11 (KHTML, like Gecko) 
Chrome/17.0.963.56 Safari/535.11",
    "Mozilla/5.0 (Macintosh; Intel Mac OS X 10_7_3) AppleWebKit/535.20 (KHTML, 
like Gecko) Chrome/19.0.1036.7 Safari/535.20",
    "Opera/9.80 (Macintosh; Intel Mac OS X 10.6.8; U; fr) Presto/2.9.168 
Version/11.52",
]

def create_headers():
    headers = dict()
    headers["User-Agent"] = random.choice(USER_AGENTS)
    headers["Referer"] = "http://www.{0}.com".format(SPIDER_NAME)
    return headers
```

8.5.3 在线 IP 代理

编写文件 proxy.py，功能是模拟使用专业在线代理中的 IP 地址，具体实现代码如下：

```
def spider_proxyip(num=10):
    try:
        url = 'http://www.网站域名.com/nt/1'
        req = requests.get(url, headers=create_headers())
        source_code = req.content
        print(source_code)
        soup = BeautifulSoup(source_code, 'lxml')
        ips = soup.findAll('tr')

        for x in range(1, len(ips)):
            ip = ips[x]
            tds = ip.findAll("td")
            proxy_host = "{0}://".format(tds[5].contents[0]) + tds[1].contents[0] 
+ ":" + tds[2].contents[0]
            proxy_temp = {tds[5].contents[0]: proxy_host}
            proxys_src.append(proxy_temp)
            if x >= num:
                break
    except Exception as e:
        print("spider_proxyip exception:")
        print(e)
```

8.6 爬虫抓取信息

本系统的核心是爬虫抓取房价信息，本节将详细讲解抓取不同类型房价信息的过程。

8.6.1 设置解析元素

编写文件 xpath.py，功能是根据要爬取的目标网站设置要抓取

扫码观看本节视频讲解

的 HTML 元素，具体实现代码如下：

```python
from lib.spider.base_spider import SPIDER_NAME, LIANJIA_SPIDER, BEIKE_SPIDER

if SPIDER_NAME == LIANJIA_SPIDER:
    ERSHOUFANG_QU_XPATH = '//*[@id="filter-options"]/dl[1]/dd/div/a'
    ERSHOUFANG_BANKUAI_XPATH = '//*[@id="filter-options"]/dl[1]/dd/div[2]/a'
    XIAOQU_QU_XPATH = '//*[@id="filter-options"]/dl[1]/dd/div/a'
    XIAOQU_BANKUAI_XPATH = '//*[@id="filter-options"]/dl[1]/dd/div[2]/a'
    DISTRICT_AREA_XPATH = '//div[3]/div[1]/dl[2]/dd/div/div[2]/a'
    CITY_DISTRICT_XPATH = '///div[3]/div[1]/dl[2]/dd/div/div/a'
elif SPIDER_NAME == BEIKE_SPIDER:
    ERSHOUFANG_QU_XPATH = '//*[@id="filter-options"]/dl[1]/dd/div/a'
    ERSHOUFANG_BANKUAI_XPATH = '//*[@id="filter-options"]/dl[1]/dd/div[2]/a'
    XIAOQU_QU_XPATH = '//*[@id="filter-options"]/dl[1]/dd/div/a'
    XIAOQU_BANKUAI_XPATH = '//*[@id="filter-options"]/dl[1]/dd/div[2]/a'
    DISTRICT_AREA_XPATH = '//div[3]/div[1]/dl[2]/dd/div/div[2]/a'
    CITY_DISTRICT_XPATH = '///div[3]/div[1]/dl[2]/dd/div/div/a'
```

8.6.2 爬取二手房信息

(1) 编写文件 ershou_spider.py 定义爬取二手房数据的爬虫派生类，具体实现流程如下：

▶ 编写函数 collect_area_ershou_data( )，功能是获取每个板块下所有的二手房信息，并且将这些信息写入 CSV 文件中保存。对应代码如下：

```python
def collect_area_ershou_data(self, city_name, area_name, fmt="csv"):
    """
      :param city_name: 城市
    :param area_name: 板块
    :param fmt: 保存文件格式
    :return: None
    """
    district_name = area_dict.get(area_name, "")
    csv_file = self.today_path + "/{0}_{1}.csv".format(district_name, area_name)
    with open(csv_file, "w") as f:
        # 开始获得需要的板块数据
        ershous = self.get_area_ershou_info(city_name, area_name)
        # 锁定,多线程读写
        if self.mutex.acquire(1):
            self.total_num += len(ershous)
            # 释放
            self.mutex.release()
        if fmt == "csv":
            for ershou in ershous:
                # print(date_string + "," + xiaoqu.text())
                f.write(self.date_string + "," + ershou.text() + "\n")
    print("Finish crawl area: " + area_name + ", save data to : " + csv_file)
```

▶ 编写函数 get_area_ershou_info( )，功能是通过爬取页面获得城市指定板块的二手房信息，对应代码如下：

```python
@staticmethod
def get_area_ershou_info(city_name, area_name):
    """
    :param city_name: 城市
    :param area_name: 板块
    :return: 二手房数据列表
    """
    total_page = 1
    district_name = area_dict.get(area_name, "")
    # 中文区县
    chinese_district = get_chinese_district(district_name)
    # 中文板块
    chinese_area = chinese_area_dict.get(area_name, "")

    ershou_list = list()
    page = 'http://{0}.{1}.com/ershoufang/{2}/'.format(city_name,
SPIDER_NAME, area_name)
    print(page)  # 打印板块页面地址
    headers = create_headers()
    response = requests.get(page, timeout=10, headers=headers)
    html = response.content
    soup = BeautifulSoup(html, "lxml")

    # 获得总的页数,通过查找总页码的元素信息
    try:
        page_box = soup.find_all('div', class_='page-box')[0]
        matches = re.search('.*"totalPage":(\d+),.*', str(page_box))
        total_page = int(matches.group(1))
    except Exception as e:
        print("\tWarning: only find one page for {0}".format(area_name))
        print(e)

    # 从第一页开始,一直遍历到最后一页
    for num in range(1, total_page + 1):
        page = 'http://{0}.{1}.com/ershoufang/{2}/pg{3}'.format(city_name,
SPIDER_NAME, area_name, num)
        print(page)  # 打印每一页的地址
        headers = create_headers()
        BaseSpider.random_delay()
        response = requests.get(page, timeout=10, headers=headers)
        html = response.content
        soup = BeautifulSoup(html, "lxml")

        # 获得有小区信息的panel
        house_elements = soup.find_all('li', class_="clear")
        for house_elem in house_elements:
            price = house_elem.find('div', class_="totalPrice")
            name = house_elem.find('div', class_='title')
            desc = house_elem.find('div', class_="houseInfo")
            pic = house_elem.find('a', class_="img").find('img', class_="lj-lazy")

            # 继续清理数据
            price = price.text.strip()
            name = name.text.replace("\n", "")
            desc = desc.text.replace("\n", "").strip()
```

```
            pic = pic.get('data-original').strip()
            # print(pic)


            # 作为对象保存
            ershou = ErShou(chinese_district, chinese_area, name, price, desc, pic)
            ershou_list.append(ershou)
    return ershou_list
```

▶ 编写函数 start(self)，功能是根据获取的城市参数来爬取这个城市的二手房信息，对应的实现代码如下：

```
def start(self):
    city = get_city()
    self.today_path = create_date_path("{0}/ershou".format(SPIDER_NAME), city, self.date_string)

    t1 = time.time()   # 开始计时

    # 获得城市有多少区列表, district: 区县
    districts = get_districts(city)
    print('City: {0}'.format(city))
    print('Districts: {0}'.format(districts))

    # 获得每个区的板块, area: 板块
    areas = list()
    for district in districts:
        areas_of_district = get_areas(city, district)
        print('{0}: Area list: {1}'.format(district, areas_of_district))
        # 用list的extend方法,L1.extend(L2),该方法将参数L2 的全部元素添加到L1 的尾部
        areas.extend(areas_of_district)
        # 使用一个字典来存储区县和板块的对应关系,如{'beicai': 'pudongxinqu', }
        for area in areas_of_district:
            area_dict[area] = district
    print("Area:", areas)
    print("District and areas:", area_dict)

    # 准备线程池用到的参数
    nones = [None for i in range(len(areas))]
    city_list = [city for i in range(len(areas))]
    args = zip(zip(city_list, areas), nones)
    # areas = areas[0: 1]    # For debugging

    # 针对每个板块写一个文件,启动一个线程来操作
    pool_size = thread_pool_size
    pool = threadpool.ThreadPool(pool_size)
    my_requests = threadpool.makeRequests(self.collect_area_ershou_data, args)
    [pool.putRequest(req) for req in my_requests]
    pool.wait()
    pool.dismissWorkers(pool_size, do_join=True)    # 完成后退出

    # 计时结束,统计结果
    t2 = time.time()
    print("Total crawl {0} areas.".format(len(areas)))
```

```
        print("Total cost {0} second to crawl {1} data items.".format(t2 - t1,
self.total_num))
```

(2) 编写文件 ershou.py，功能是爬取指定城市的二手房信息，具体实现代码如下：

```
from lib.spider.ershou_spider import *

if __name__ == "__main__":
    spider = ErShouSpider(SPIDER_NAME)
    spider.start()
```

执行文件 ershou.py 后，会先提示用户选择一个要抓取的城市：

```
Today date is: 20190212
Target site is ke.com
Wait for your choice.
Which city do you want to crawl?
bj: 北京, cd: 成都, cq: 重庆, cs: 长沙
dg: 东莞, dl: 大连, fs: 佛山, gz: 广州
hz: 杭州, hf: 合肥, jn: 济南, nj: 南京
qd: 青岛, sh: 上海, sz: 深圳, su: 苏州
sy: 沈阳, tj: 天津, wh: 武汉, xm: 厦门
yt: 烟台,
```

输入一个城市的两个字母标识，如输入 bj 并按回车键后，会抓取当天北京市的二手房信息，并将抓取到的信息保存到 CSV 文件中，如图 8-2 所示。

图 8-2　抓取到的二手房信息被保存到 CSV 文件中

8.6.3　爬取楼盘信息

(1) 编写文件 loupan_spider.py 定义爬取楼盘数据的爬虫派生类，具体实现流程如下：

- 编写函数collect_city_loupan_data( )，功能是将指定城市的新房楼盘数据存储下来，并将抓取的信息默认存储到CSV文件中。对应的实现代码如下：

```python
def collect_city_loupan_data(self, city_name, fmt="csv"):
    """
    :param city_name: 城市
    :param fmt: 保存文件格式
    :return: None
    """
    csv_file = self.today_path + "/{0}.csv".format(city_name)
    with open(csv_file, "w") as f:
        # 开始获得需要的板块数据
        loupans = self.get_loupan_info(city_name)
        self.total_num = len(loupans)
        if fmt == "csv":
            for loupan in loupans:
                f.write(self.date_string + "," + loupan.text() + "\n")
    print("Finish crawl: " + city_name + ", save data to : " + csv_file)
```

- 编写函数get_loupan_info( )，功能是爬取指定目标城市的新房楼盘信息，对应代码如下：

```python
@staticmethod
def get_loupan_info(city_name):
    """
    :param city_name: 城市
    :return: 新房楼盘信息列表
    """
    total_page = 1
    loupan_list = list()
    page = 'http://{0}.fang.{1}.com/loupan/'.format(city_name, SPIDER_NAME)
    print(page)
    headers = create_headers()
    response = requests.get(page, timeout=10, headers=headers)
    html = response.content
    soup = BeautifulSoup(html, "lxml")

    # 获得总的页数
    try:
        page_box = soup.find_all('div', class_='page-box')[0]
        matches = re.search('.*data-total-count="(\d+)".*', str(page_box))
        total_page = int(math.ceil(int(matches.group(1)) / 10))
    except Exception as e:
        print("\tWarning: only find one page for {0}".format(city_name))
        print(e)

    print(total_page)
    # 从第一页开始,一直遍历到最后一页
    headers = create_headers()
    for i in range(1, total_page + 1):
        page = 'http://{0}.fang.{1}.com/loupan/pg{2}'.format(city_name, SPIDER_NAME, i)
        print(page)
        BaseSpider.random_delay()
```

```python
        response = requests.get(page, timeout=10, headers=headers)
        html = response.content
        soup = BeautifulSoup(html, "lxml")

        # 获得有小区信息的panel
        house_elements = soup.find_all('li', class_="resblock-list")
        for house_elem in house_elements:
            price = house_elem.find('span', class_="number")
            total = house_elem.find('div', class_="second")
            loupan = house_elem.find('a', class_='name')

            # 继续清理数据
            try:
                price = price.text.strip()
            except Exception as e:
                price = '0'

            loupan = loupan.text.replace("\n", "")

            try:
                total = total.text.strip().replace(u'总价', '')
                total = total.replace(u'/套起', '')
            except Exception as e:
                total = '0'

            print("{0} {1} {2} ".format(
                loupan, price, total))

            # 作为对象保存
            loupan = LouPan(loupan, price, total)
            loupan_list.append(loupan)
    return loupan_list
```

- 编写函数 start(self)，功能是根据获取的城市参数来爬取这个城市的二手房信息，对应的实现代码如下：

```
def start(self):
    city = get_city()
    print('Today date is: %s' % self.date_string)
    self.today_path = create_date_path("{0}/loupan".format(SPIDER_NAME),
city, self.date_string)

    t1 = time.time()  # 开始计时
    self.collect_city_loupan_data(city)
    t2 = time.time()  # 计时结束,统计结果

    print("Total crawl {0} loupan.".format(self.total_num))
    print("Total cost {0} second ".format(t2 - t1))
```

(2) 编写文件 loupan.py，功能是爬取指定城市的新房楼盘信息，具体实现代码如下：

```
from lib.spider.loupan_spider import *

if __name__ == "__main__":
    spider = LouPanBaseSpider(SPIDER_NAME)
    spider.start()
```

执行文件 loupan.py 后，会先提示用户选择一个要抓取的城市：

```
Today date is: 20190212
Target site is ke.com
Wait for your choice.
Which city do you want to crawl?
bj: 北京, cd: 成都, cq: 重庆, cs: 长沙
dg: 东莞, dl: 大连, fs: 佛山, gz: 广州
hz: 杭州, hf: 合肥, jn: 济南, nj: 南京
qd: 青岛, sh: 上海, sz: 深圳, su: 苏州
sy: 沈阳, tj: 天津, wh: 武汉, xm: 厦门
yt: 烟台,
```

输入一个城市的两个字母标识，如输入 jn 并按回车键后，会抓取当天济南市的新房楼盘信息，并将抓取到的信息保存到 CSV 文件中，如图 8-3 所示。

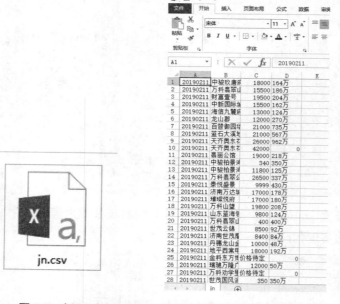

图 8-3　抓取到的新房楼盘信息被保存到 CSV 文件中

8.6.4　爬取小区信息

(1) 编写文件 xiaoqu_spider.py 定义爬取小区数据的爬虫派生类，具体实现流程如下。

▶ 编写函数 collect_area_xiaoqu_data( )，功能是获取每个板块下的所有小区的信息，并且将这些信息写入 CSV 文件中进行保存。对应代码如下：

```
def collect_area_xiaoqu_data(self, city_name, area_name, fmt="csv"):
    """
    :param city_name: 城市
    :param area_name: 板块
    :param fmt: 保存文件格式
    :return: None
```

```python
        """
        district_name = area_dict.get(area_name, "")
        csv_file = self.today_path + "/{0}_{1}.csv".format(district_name, area_name)
        with open(csv_file, "w") as f:
            # 开始获得需要的板块数据
            xqs = self.get_xiaoqu_info(city_name, area_name)
            # 锁定
            if self.mutex.acquire(1):
                self.total_num += len(xqs)
                # 释放
                self.mutex.release()
            if fmt == "csv":
                for xiaoqu in xqs:
                    f.write(self.date_string + "," + xiaoqu.text() + "\n")
        print("Finish crawl area: " + area_name + ", save data to : " + csv_file)
        logger.info("Finish crawl area: " + area_name + ", save data to : " + csv_file)
```

▶ 编写函数 get_xiaoqu_info( )，功能是获取指定小区的详细信息，对应代码如下：

```python
@staticmethod
def get_xiaoqu_info(city, area):
    total_page = 1
    district = area_dict.get(area, "")
    chinese_district = get_chinese_district(district)
    chinese_area = chinese_area_dict.get(area, "")
    xiaoqu_list = list()
    page = 'http://{0}.{1}.com/xiaoqu/{2}/'.format(city, SPIDER_NAME, area)
    print(page)
    logger.info(page)

    headers = create_headers()
    response = requests.get(page, timeout=10, headers=headers)
    html = response.content
    soup = BeautifulSoup(html, "lxml")

    # 获得总的页数
    try:
        page_box = soup.find_all('div', class_='page-box')[0]
        matches = re.search('.*"totalPage":(\d+),.*', str(page_box))
        total_page = int(matches.group(1))
    except Exception as e:
        print("\tWarning: only find one page for {0}".format(area))
        print(e)

    # 从第一页开始,一直遍历到最后一页
    for i in range(1, total_page + 1):
        headers = create_headers()
        page = 'http://{0}.{1}.com/xiaoqu/{2}/pg{3}'.format(city, SPIDER_NAME, area, i)
        print(page)    # 打印板块页面地址
        BaseSpider.random_delay()
        response = requests.get(page, timeout=10, headers=headers)
        html = response.content
```

```python
        soup = BeautifulSoup(html, "lxml")

        # 获得有小区信息的 panel
        house_elems = soup.find_all('li', class_="xiaoquListItem")
        for house_elem in house_elems:
            price = house_elem.find('div', class_="totalPrice")
            name = house_elem.find('div', class_='title')
            on_sale = house_elem.find('div', class_="xiaoquListItemSellCount")

            # 继续清理数据
            price = price.text.strip()
            name = name.text.replace("\n", "")
            on_sale = on_sale.text.replace("\n", "").strip()

            # 作为对象保存
            xiaoqu = XiaoQu(chinese_district, chinese_area, name, price, on_sale)
            xiaoqu_list.append(xiaoqu)
    return xiaoqu_list
```

- 编写函数 start(self)，功能是根据获取的城市参数来爬取这个城市的小区信息，对应的实现代码如下：

```
def start(self):
    city = get_city()
    self.today_path = create_date_path("{0}/xiaoqu".format(SPIDER_NAME), city, self.date_string)
    t1 = time.time()   # 开始计时

    # 获得城市有多少区列表, district: 区县
    districts = get_districts(city)
    print('City: {0}'.format(city))
    print('Districts: {0}'.format(districts))

    # 获得每个区的板块, area: 板块
    areas = list()
    for district in districts:
        areas_of_district = get_areas(city, district)
        print('{0}: Area list: {1}'.format(district, areas_of_district))
        # 用list的extend方法,L1.extend(L2)，该方法将参数L2的全部元素添加到L1的尾部
        areas.extend(areas_of_district)
        # 使用一个字典来存储区县和板块的对应关系,如{'beicai': 'pudongxinqu', }
        for area in areas_of_district:
            area_dict[area] = district
    print("Area:", areas)
    print("District and areas:", area_dict)

    # 准备线程池用到的参数
    nones = [None for i in range(len(areas))]
    city_list = [city for i in range(len(areas))]
    args = zip(zip(city_list, areas), nones)
    # areas = areas[0: 1]

    # 针对每个板块写一个文件,启动一个线程来操作
    pool_size = thread_pool_size
    pool = threadpool.ThreadPool(pool_size)
```

```python
        my_requests = threadpool.makeRequests(self.collect_area_xiaoqu_data, args)
        [pool.putRequest(req) for req in my_requests]
        pool.wait()
        pool.dismissWorkers(pool_size, do_join=True)  # 完成后退出

        # 计时结束,统计结果
        t2 = time.time()
        print("Total crawl {0} areas.".format(len(areas)))
        print("Total cost {0} second to crawl {1} data items.".format(t2 - t1,
self.total_num))
```

(2) 编写文件 xiaoqu.py，功能是爬取指定城市的小区信息，具体实现代码如下：

```python
from lib.spider.xiaoqu_spider import *

if __name__ == "__main__":
    spider = XiaoQuBaseSpider(SPIDER_NAME)
    spider.start()
```

执行文件 xiaoqu.py 后，会先提示用户选择一个要爬取的城市：

```
Today date is: 20190212
Target site is ke.com
Wait for your choice.
Which city do you want to crawl?
bj: 北京,  cd: 成都,  cq: 重庆,  cs: 长沙
dg: 东莞,  dl: 大连,  fs: 佛山,  gz: 广州
hz: 杭州,  hf: 合肥,  jn: 济南,  nj: 南京
qd: 青岛,  sh: 上海,  sz: 深圳,  su: 苏州
sy: 沈阳,  tj: 天津,  wh: 武汉,  xm: 厦门
yt: 烟台,
```

输入一个城市的两个字母标识，如输入 jn 并按回车键后，会抓取当天济南市的小区信息，并将爬取到的信息保存到 CSV 文件中，如图 8-4 所示。

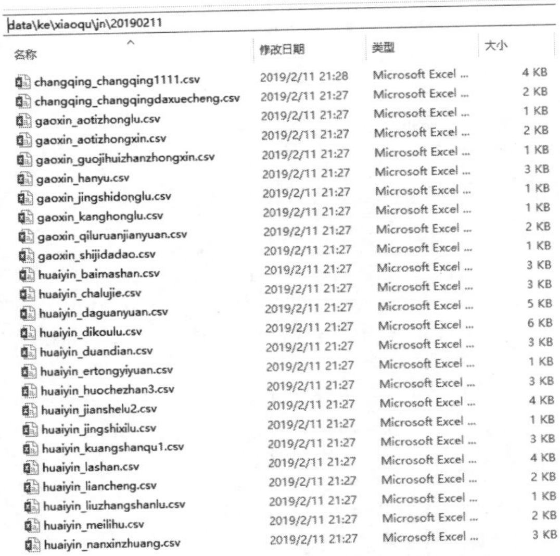

图 8-4　爬取到的小区信息被保存到 CSV 文件中

8.6.5 抓取租房信息

(1) 编写文件 zufang_spider.py 定义爬取租房数据的爬虫派生类，具体实现流程如下：

▶ 编写函数 collect_area_zufang_data( )，功能是获取每个板块下的所有出租房的信息，并且将这些信息写入 CSV 文件中进行保存。对应的实现代码如下：

```python
def collect_area_zufang_data(self, city_name, area_name, fmt="csv"):
    """
    :param city_name: 城市
    :param area_name: 板块
    :param fmt: 保存文件格式
    :return: None
    """
    district_name = area_dict.get(area_name, "")
    csv_file = self.today_path + "/{0}_{1}.csv".format(district_name, area_name)
    with open(csv_file, "w") as f:
        # 开始获得需要的板块数据
        zufangs = self.get_area_zufang_info(city_name, area_name)
        # 锁定
        if self.mutex.acquire(1):
            self.total_num += len(zufangs)
            # 释放
            self.mutex.release()
        if fmt == "csv":
            for zufang in zufangs:
                f.write(self.date_string + "," + zufang.text() + "\n")
    print("Finish crawl area: " + area_name + ", save data to : " + csv_file)
```

▶ 编写函数 get_area_zufang_info( )，功能是获取指定城市指定板块的租房信息。对应的实现代码如下：

```python
@staticmethod
def get_area_zufang_info(city_name, area_name):
    matches = None
    """
    :param city_name: 城市
    :param area_name: 板块
    :return: 出租房信息列表
    """
    total_page = 1
    district_name = area_dict.get(area_name, "")
    chinese_district = get_chinese_district(district_name)
    chinese_area = chinese_area_dict.get(area_name, "")
    zufang_list = list()
    page = 'http://{0}.{1}.com/zufang/{2}/'.format(city_name, SPIDER_NAME, area_name)
    print(page)

    headers = create_headers()
    response = requests.get(page, timeout=10, headers=headers)
    html = response.content
    soup = BeautifulSoup(html, "lxml")
```

```python
        # 获得总的页数
        try:
            if SPIDER_NAME == "lianjia":
                page_box = soup.find_all('div', class_='page-box')[0]
                matches = re.search('.*"totalPage":(\d+),.*', str(page_box))
            elif SPIDER_NAME == "ke":
                page_box = soup.find_all('div', class_='content__pg')[0]
                # print(page_box)
                matches = re.search('.*data-totalpage="(\d+)".*', str(page_box))
            total_page = int(matches.group(1))
            # print(total_page)
        except Exception as e:
            print("\tWarning: only find one page for {0}".format(area_name))
            print(e)

        # 从第一页开始，一直遍历到最后一页
        headers = create_headers()
        for num in range(1, total_page + 1):
            page = 'http://{0}.{1}.com/zufang/{2}/pg{3}'.format(city_name,
SPIDER_NAME, area_name, num)
            print(page)
            BaseSpider.random_delay()
            response = requests.get(page, timeout=10, headers=headers)
            html = response.content
            soup = BeautifulSoup(html, "lxml")

            # 获得有小区信息的panel
            if SPIDER_NAME == "lianjia":
                ul_element = soup.find('ul', class_="house-lst")
                house_elements = ul_element.find_all('li')
            else:
                ul_element = soup.find('div', class_="content__list")
                house_elements = ul_element.find_all('div', class_=
"content__list--item")

            if len(house_elements) == 0:
                continue
            # else:
            #     print(len(house_elements))

            for house_elem in house_elements:
                if SPIDER_NAME == "lianjia":
                    price = house_elem.find('span', class_="num")
                    xiaoqu = house_elem.find('span', class_='region')
                    layout = house_elem.find('span', class_="zone")
                    size = house_elem.find('span', class_="meters")
                else:
                    price = house_elem.find('span', class_=
"content__list--item-price")
                    desc1 = house_elem.find('p', class_=
"content__list--item--title")
                    desc2 = house_elem.find('p', class_=
"content__list--item--des")
```

```python
            try:
                if SPIDER_NAME == "lianjia":
                    price = price.text.strip()
                    xiaoqu = xiaoqu.text.strip().replace("\n", "")
                    layout = layout.text.strip()
                    size = size.text.strip()
                else:
                    # 继续清理数据
                    price = price.text.strip().replace(" ", "").replace("元/月", "")
                    # print(price)
                    desc1 = desc1.text.strip().replace("\n", "")
                    desc2 = desc2.text.strip().replace("\n", "").replace(" ", "")
                    # print(desc1)

                    infos = desc1.split(' ')
                    xiaoqu = infos[0]
                    layout = infos[1]
                    descs = desc2.split('/')
                    # print(descs[1])
                    size = descs[1].replace("m²", "平米")

                # print("{0} {1} {2} {3} {4} {5} {6}".format(
                #     chinese_district, chinese_area, xiaoqu, layout, size, price))

                # 作为对象保存
                zufang = ZuFang(chinese_district, chinese_area, xiaoqu, layout, size, price)
                zufang_list.append(zufang)
            except Exception as e:
                print("=" * 20 + " page no data")
                print(e)
                print(page)
                print("=" * 20)
    return zufang_list
```

▶ 编写函数 start(self)，功能是根据获取的城市参数来爬取这个城市的二手房信息，对应的实现代码如下：

```python
def start(self):
    city = get_city()
    self.today_path = create_date_path("{0}/zufang".format(SPIDER_NAME), city, self.date_string)
    # collect_area_zufang('sh', 'beicai')  # For debugging, keep it here
    t1 = time.time()  # 开始计时

    # 获得城市有多少区列表, district: 区县
    districts = get_districts(city)
    print('City: {0}'.format(city))
    print('Districts: {0}'.format(districts))

    # 获得每个区的板块, area: 板块
    areas = list()
    for district in districts:
        areas_of_district = get_areas(city, district)
```

```
        print('{0}: Area list: {1}'.format(district, areas_of_district))
        # 用list的extend方法，L1.extend(L2)，该方法将参数L2的全部元素添加到L1的尾部
        areas.extend(areas_of_district)
        # 使用一个字典来存储区县和板块的对应关系，如{'beicai': 'pudongxinqu', }
        for area in areas_of_district:
            area_dict[area] = district
    print("Area:", areas)
    print("District and areas:", area_dict)

    # 准备线程池用到的参数
    nones = [None for i in range(len(areas))]
    city_list = [city for i in range(len(areas))]
    args = zip(zip(city_list, areas), nones)
    # areas = areas[0: 1]

    # 针对每个板块写一个文件,启动一个线程来操作
    pool_size = thread_pool_size
    pool = threadpool.ThreadPool(pool_size)
    my_requests = threadpool.makeRequests(self.collect_area_zufang_data, args)
    [pool.putRequest(req) for req in my_requests]
    pool.wait()
    pool.dismissWorkers(pool_size, do_join=True)  # 完成后退出

    # 计时结束,统计结果
    t2 = time.time()
    print("Total crawl {0} areas.".format(len(areas)))
    print("Total cost {0} second to crawl {1} data items.".format(t2 - t1, self.total_num))
```

(2) 编写文件 zufang.py，功能是爬取指定城市的租房信息，具体实现代码如下：

```
from lib.spider.zufang_spider import *

if __name__ == "__main__":
    spider = ZuFangBaseSpider(SPIDER_NAME)
    spider.start()
```

执行文件 zufang.py 后，会先提示用户选择一个要抓取的城市：

```
Today date is: 20190212
Target site is ke.com
Wait for your choice.
Which city do you want to crawl?
bj: 北京, cd: 成都, cq: 重庆, cs: 长沙
dg: 东莞, dl: 大连, fs: 佛山, gz: 广州
hz: 杭州, hf: 合肥, jn: 济南, nj: 南京
qd: 青岛, sh: 上海, sz: 深圳, su: 苏州
sy: 沈阳, tj: 天津, wh: 武汉, xm: 厦门
yt: 烟台,
```

输入一个城市的两个字母标识，如输入 jn 并按回车键后，会抓取当天济南市的租房信息，并将抓取到的信息保存到 CSV 文件中，如图 8-5 所示。

```
gaoxin_aotizhongxin.csv        2019/2/11 20:53    Microsoft Excel ...    7 KB
gaoxin_hanyu.csv               2019/2/11 20:54    Microsoft Excel ...    0 KB
gaoxin_jingshidonglu.csv       2019/2/11 20:57    Microsoft Excel ...   18 KB
gaoxin_kanghonglu.csv          2019/2/11 20:57    Microsoft Excel ...    5 KB
gaoxin_qiluruanjianyuan.csv    2019/2/11 20:55    Microsoft Excel ...    0 KB
huaiyin_baimashan.csv          2019/2/11 20:57    Microsoft Excel ...    0 KB
huaiyin_chalujie.csv           2019/2/11 20:57    Microsoft Excel ...    4 KB
huaiyin_daguanyuan.csv         2019/2/11 20:56    Microsoft Excel ...    0 KB
huaiyin_dikoulu.csv            2019/2/11 20:56    Microsoft Excel ...    0 KB
huaiyin_duandian.csv           2019/2/11 20:56    Microsoft Excel ...    0 KB
huaiyin_ertongyiyuan.csv       2019/2/11 20:56    Microsoft Excel ...    0 KB
huaiyin_huochezhan3.csv        2019/2/11 20:56    Microsoft Excel ...    0 KB
huaiyin_jianshelu2.csv         2019/2/11 20:57    Microsoft Excel ...    0 KB
huaiyin_jingshixilu.csv        2019/2/11 20:57    Microsoft Excel ...    1 KB
huaiyin_kuangshanqu1.csv       2019/2/11 20:56    Microsoft Excel ...    0 KB
huaiyin_lashan.csv             2019/2/11 20:56    Microsoft Excel ...    0 KB
huaiyin_liancheng.csv          2019/2/11 20:56    Microsoft Excel ...    0 KB
huaiyin_liuzhangshanlu.csv     2019/2/11 20:57    Microsoft Excel ...    1 KB
huaiyin_meilihu.csv            2019/2/11 20:57    Microsoft Excel ...    0 KB
huaiyin_nanxinzhuang.csv       2019/2/11 20:57    Microsoft Excel ...    0 KB
huaiyin_quanjingwolongqu.csv   2019/2/11 20:57    Microsoft Excel ...    0 KB
huaiyin_shengliyiyuan.csv      2019/2/11 20:57    Microsoft Excel ...    0 KB
huaiyin_shigou.csv             2019/2/11 20:57    Microsoft Excel ...    0 KB
huaiyin_tianqiaogongyeyuan.csv 2019/2/11 20:57    Microsoft Excel ...    1 KB
huaiyin_wangguanzhuang.csv     2019/2/11 20:57    Microsoft Excel ...    0 KB
huaiyin_xikezhan.csv           2019/2/11 20:57    Microsoft Excel ...    0 KB
```

图 8-5　抓取到的租房信息被保存到 CSV 文件中

8.7　数据可视化

在抓取到房价数据后,可以将 CSV 文件实现可视化分析。但是为了更加方便地操作,可以将抓取的数据保存到数据库中,然后提取数据库中的数据进行数据分析。在本节的内容中,将详细讲解将数据保存到数据库并进行数据分析的过程。

扫码观看本节视频讲解

8.7.1　爬取数据并保存到数据库

编写文件 xiaoqu_to_db.py,功能是抓取指定城市的小区房价数据保存到数据库中,可以选择存储的数据库类型有 MySQL、MongoDB、JSON、CSV 和 Excel,默认的存储方式是 MySQL。文件 xiaoqu_to_db.py 的具体实现流程如下。

(1) 创建提示语句,询问用户将要抓取的目标城市,对应的实现代码如下:

```python
pymysql.install_as_MySQLdb()
def create_prompt_text():
    city_info = list()
    num = 0
    for en_name, ch_name in cities.items():
        num += 1
        city_info.append(en_name)
        city_info.append(": ")
        city_info.append(ch_name)
        if num % 4 == 0:
            city_info.append("\n")
        else:
            city_info.append(", ")
    return 'Which city data do you want to save ?\n' + ''.join(city_info)
```

(2) 设置数据库类型，根据不同的存储类型执行对应的写入操作，对应的实现代码如下：

```python
if __name__ == '__main__':
    # 设置目标数据库
    ####################################
    # mysql/mongodb/excel/json/csv
    database = "mysql"
    # database = "mongodb"
    # database = "excel"
    # database = "json"
    # database = "csv"
    ####################################
    db = None
    collection = None
    workbook = None
    csv_file = None
    datas = list()

    if database == "mysql":
        import records
        db = records.Database('mysql://root:66688888@localhost/lianjia?charset=utf8', encoding='utf-8')
    elif database == "mongodb":
        from pymongo import MongoClient
        conn = MongoClient('localhost', 27017)
        db = conn.lianjia  # 连接 lianjia 数据库，没有则自动创建
        collection = db.xiaoqu  # 使用 xiaoqu 集合，没有则自动创建
    elif database == "excel":
        import xlsxwriter
        workbook = xlsxwriter.Workbook('xiaoqu.xlsx')
        worksheet = workbook.add_worksheet()
    elif database == "json":
        import json
    elif database == "csv":
        csv_file = open("xiaoqu.csv", "w")
        line = "{0};{1};{2};{3};{4};{5};{6}\n".format('city_ch', 'date', 'district', 'area', 'xiaoqu', 'price', 'sale')
        csv_file.write(line)
```

(3) 准备日期信息，将爬到的数据保存到对应日期的相关文件夹下，对应的实现代码如下：

```python
    city = get_city()
    date = get_date_string()
    # 获得 csv 文件路径
    # date = "20180331"    # 指定采集数据的日期
    # city = "sh"          # 指定采集数据的城市
    city_ch = get_chinese_city(city)
    csv_dir = "{0}/{1}/xiaoqu/{2}/{3}".format(DATA_PATH, SPIDER_NAME, city, date)

    files = list()
    if not os.path.exists(csv_dir):
        print("{0} does not exist.".format(csv_dir))
        print("Please run 'python xiaoqu.py' firstly.")
```

```python
        print("Bye.")
        exit(0)
    else:
        print('OK, start to process ' + get_chinese_city(city))
    for csv in os.listdir(csv_dir):
        data_csv = csv_dir + "/" + csv
        # print(data_csv)
        files.append(data_csv)
```

(4) 清理数据，删除没有房源信息的小区，对应的实现代码如下：

```python
# 清理数据
count = 0
row = 0
col = 0
for csv in files:
    with open(csv, 'r') as f:
        for line in f:
            count += 1
            text = line.strip()
            try:
                # 如果小区名里面没有逗号，那么总共是 6 项
                if text.count(',') == 5:
                    date, district, area, xiaoqu, price, sale = text.split(',')
                elif text.count(',') < 5:
                    continue
                else:
                    fields = text.split(',')
                    date = fields[0]
                    district = fields[1]
                    area = fields[2]
                    xiaoqu = ','.join(fields[3:-2])
                    price = fields[-2]
                    sale = fields[-1]
            except Exception as e:
                print(text)
                print(e)
                continue
            sale = sale.replace(r'套在售二手房', '')
            price = price.replace(r'暂无', '0')
            price = price.replace(r'元/m2', '')
            price = int(price)
            sale = int(sale)
            print("{0} {1} {2} {3} {4} {5}".format(date, district, area, xiaoqu, price, sale))
```

(5) 将爬取到的房价数据添加到数据库中或 JSON、Excel、CSV 文件中，对应的实现代码如下：

```python
            # 写入mysql数据库
            if database == "mysql":
                db.query('INSERT INTO xiaoqu (city, date, district, area, xiaoqu, price, sale) '
                         'VALUES (:city, :date, :district, :area, :xiaoqu, :price, :sale)',
```

```python
                        city=city_ch, date=date, district=district, area=area,
xiaoqu=xiaoqu, price=price,
                        sale=sale)
                # 写入 mongodb 数据库
                elif database == "mongodb":
                    data = dict(city=city_ch, date=date, district=district,
area=area, xiaoqu=xiaoqu, price=price,
                        sale=sale)
                    collection.insert(data)
                elif database == "excel":
                    if not PYTHON_3:
                        worksheet.write_string(row, col, city_ch)
                        worksheet.write_string(row, col + 1, date)
                        worksheet.write_string(row, col + 2, district)
                        worksheet.write_string(row, col + 3, area)
                        worksheet.write_string(row, col + 4, xiaoqu)
                        worksheet.write_number(row, col + 5, price)
                        worksheet.write_number(row, col + 6, sale)
                    else:
                        worksheet.write_string(row, col, city_ch)
                        worksheet.write_string(row, col + 1, date)
                        worksheet.write_string(row, col + 2, district)
                        worksheet.write_string(row, col + 3, area)
                        worksheet.write_string(row, col + 4, xiaoqu)
                        worksheet.write_number(row, col + 5, price)
                        worksheet.write_number(row, col + 6, sale)
                    row += 1
                elif database == "json":
                    data = dict(city=city_ch, date=date, district=district,
area=area, xiaoqu=xiaoqu, price=price,
                        sale=sale)
                    datas.append(data)
                elif database == "csv":
                    line = "{0};{1};{2};{3};{4};{5};{6}\n".format(city_ch, date,
district, area, xiaoqu, price, sale)
                    csv_file.write(line)

    # 写入,并且关闭句柄
    if database == "excel":
        workbook.close()
    elif database == "json":
        json.dump(datas, open('xiaoqu.json', 'w'), ensure_ascii=False,
indent=2)
    elif database == "csv":
        csv_file.close()

    print("Total write {0} items to database.".format(count))
```

执行后会提示用户选择一个目标城市：

```
Wait for your choice.
Which city do you want to crawl?
bj: 北京, cd: 成都, cq: 重庆, cs: 长沙
dg: 东莞, dl: 大连, fs: 佛山, gz: 广州
hz: 杭州, hf: 合肥, jn: 济南, nj: 南京
```

qd: 青岛，sh: 上海，sz: 深圳，su: 苏州
sy: 沈阳，tj: 天津，wh: 武汉，xm: 厦门
yt: 烟台，

假设输入 jn 并按回车键，则会将济南市的小区信息保存到数据库中。因为在上述代码中设置的默认存储方式是 MySQL，所以会将抓取到的数据保存到 MySQL 数据库中，如图 8-6 所示。

图 8-6　将数据保存到 MySQL 数据库中

> **注　意**
>
> MySQL 数据库的数据结构，通过导入源码目录中的 SQL 文件 lianjia_xiaoqu.sql 创建。

8.7.2　可视化济南市房价最贵的 4 个小区

编写文件 pricetubiao.py，功能是提取分析 MySQL 数据库中的数据，可视化展示当日济南市房价最贵的 4 个小区。文件 pricetubiao.py 的具体实现代码如下：

```
import pymysql
from pylab import *
mpl.rcParams["font.sans-serif"] = ["SimHei"]
mpl.rcParams["axes.unicode_minus"] = False

##获取一个数据库连接,注意如果是 UTF-8 类型的,需要指定数据库
db = pymysql.connect(host="localhost", user='root', passwd="66688888",
port=3306, db="lianjia", charset='utf8')
cursor = db.cursor()   # 获取一个游标
sql = "select xiaoqu,price from xiaoqu where price!=0 order by price desc LIMIT 4"
cursor.execute(sql)
result = cursor.fetchall()   # result 为元组
```

```
# 将元组数据存进列表中
xiaoqu = []
price = []
for x in result:
    xiaoqu.append(x[0])
    price.append(x[1])

# 直方图
plt.bar(range(len(price)), price, color='steelblue', tick_label=xiaoqu)
plt.xlabel("小区名")
plt.ylabel("价格")
plt.title("济南房价 Top 4 小区")
for x, y in enumerate(price):
    plt.text(x - 0.1, y + 1, '%s' % y)
plt.show()
cursor.close()    # 关闭游标
db.close()        # 关闭数据库
```

执行后的效果如图 8-7 所示。

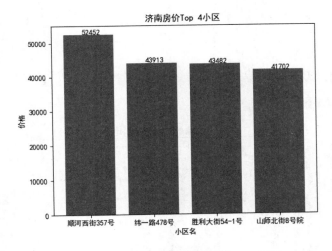

图 8-7　执行后的效果

8.7.3　可视化济南市主要地区的房价均价

编写文件 gequ.py，功能是提取分析 MySQL 数据库中的数据，可视化展示当日济南市主要行政区的房价均价。文件 gequ.py 的具体实现代码如下：

```
import pymysql
from pylab import *
mpl.rcParams["font.sans-serif"] = ["SimHei"]
mpl.rcParams["axes.unicode_minus"] = False
plt.figure(figsize=(10, 6))
##获取一个数据库连接,注意如果是UTF-8类型的,需要指定数据库
db = pymysql.connect(host="localhost", user='root', passwd="66688888",
port=3306, db="lianjia", charset='utf8')
cursor = db.cursor()    # 获取一个游标
```

```python
sql = "select district,avg(price) as avgsprice from xiaoqu where price!=0 group by district"

cursor.execute(sql)
result = cursor.fetchall()    # result 为元组

# 将元组数据存进列表中
district = []
avgsprice = []
for x in result:
    district.append(x[0])
    avgsprice.append(x[1])

# 直方图
plt.bar(range(len(avgsprice)), avgsprice, color='steelblue',
tick_label=district)
plt.xlabel("行政区")
plt.ylabel("平均价格")
plt.title("济南市主要行政区房价均价")
for x, y in enumerate(avgsprice):
    plt.text(x - 0.5, y + 2, '%s' % y)
plt.show()
cursor.close()    # 关闭游标
db.close()    # 关闭数据库
```

执行后的效果如图 8-8 所示。

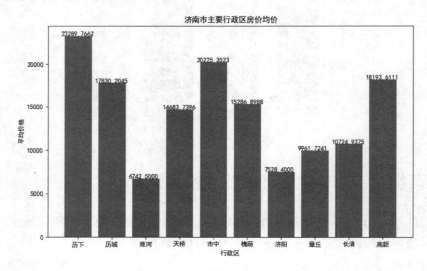

图 8-8　执行后的效果

8.7.4　可视化济南市主要地区的房源数量

编写文件 fangyuanshuliang.py，功能是提取分析 MySQL 数据库中的数据，可视化展示当日济南市主要行政区的房源数量。文件 fangyuanshuliang.py 的具体实现代码如下：

```
##获取一个数据库连接,注意如果是UTF-8类型的,需要指定数据库
db = pymysql.connect(host="localhost", user='root', passwd="66688888",
port=3306, db="lianjia", charset='utf8')
cursor = db.cursor()  # 获取一个游标
sql = "SELECT district,sum(sale) as bili FROM xiaoqu where price!=0 and sale>=1
group by district"
cursor.execute(sql)
result = cursor.fetchall()  # result 为元组

# 将元组数据存进列表中
district = []
bili = []


for x in result:
    district.append(x[0])
    bili.append(x[1])


print(district)
print(bili)

# 直方图
plt.bar(range(len(bili)), bili, color='steelblue', tick_label=district)
plt.xlabel("行政区")
plt.ylabel("房源数量")
plt.title("济南市主要行政区房源数量(套)")
for x, y in enumerate(bili):
    plt.text(x - 0.2, y + 100, '%s' % y)
plt.show()

cursor.close()   # 关闭游标
db.close()   # 关闭数据库
```

执行后的效果如图 8-9 所示。

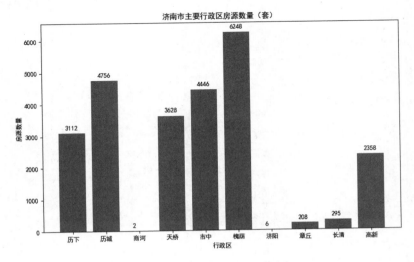

图 8-9　执行后的效果

8.7.5　可视化济南市各区的房源数量所占百分比

编写文件 bing.py，功能是提取分析 MySQL 数据库中的数据，可视化展示当日济南市主要行政区的房源数量所占百分比。为了使饼形图的界面美观，只是切片选取了济南市 7 个区的数据。文件 bing.py 的具体实现代码如下：

```python
import pymysql
from pylab import *
mpl.rcParams["font.sans-serif"] = ["SimHei"]
mpl.rcParams["axes.unicode_minus"] = False
plt.figure(figsize=(9, 7))
##获取一个数据库连接,注意如果是UTF-8类型的,需要指定数据库
db = pymysql.connect(host="localhost", user='root', passwd="66688888",
port=3306, db="lianjia", charset='utf8')
cursor = db.cursor()   # 获取一个游标

sql = "select district,sum(sale) as quzongji from xiaoqu where price!=0 group by district"
cursor.execute(sql)
result = cursor.fetchall()   # result 为元组
quzongji = []
district = []
for x in result:
    district.append(x[0])
    quzongji.append(x[1])

print(district)
print(quzongji)

sql1 = "select district,sum(sale) as quanbu from xiaoqu where price!=0"
cursor.execute(sql1)
result = cursor.fetchall()   # result 为元组

# 将元组数据存进列表中
quanbu = []
for x in result:
    quanbu.append(x[1])
print(quanbu)

import numpy as np

a = np.array(quzongji)
c = (a / quanbu)*100
print(c)


matplotlib.rcParams['font.sans-serif'] = ['SimHei']
matplotlib.rcParams['axes.unicode_minus'] = False

label_list = district[:7]      # 各部分标签
size = c[:7]       # 各部分大小
```

```
color = ["red", "green", "blue", "cyan", "magenta", "yellow", "black"]      # 各
部分颜色
explode = [0, 0, 0.2, 0, 0, 0, 0.2]      # 各部分突出值
patches, l_text, p_text = plt.pie(size, explode=explode, colors=color,
labels=label_list, labeldistance=1.1, autopct="%1.2f%%", shadow=False,
startangle=90, pctdistance=0.7)
plt.axis("equal")       # 设置横轴和纵轴大小相等,这样饼才是圆的

plt.legend()
plt.show()

cursor.close()   # 关闭游标
db.close()   # 关闭数据库
```

执行后的效果如图 8-10 所示。

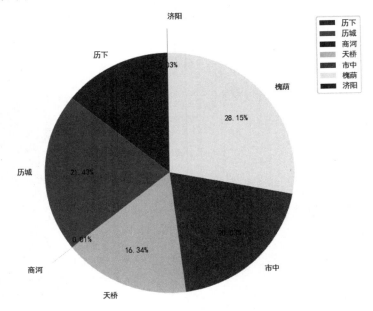

图 8-10 执行后的效果

第 9 章

商业应用：交通数据可视化

城市交通指挥中心既是城市交通管理部门的警力指挥调度中心，同时又是交通管理业务的技术支持和信息发布中心，有大量的交通管理实时和历史数据在这里汇总、存储。基于城市交通数据打造一个数据可视化平台有着十分重要的作用，这样可以更加高效、有针对性地管理城市的交通状况。

9.1 系统架构分析

本项目的交通数据被保存在 CSV 文件 bikes.csv 中,在里面保存了蒙特利尔的一些骑自行车数据,详细记录了每天在蒙特利尔 7 条不同的道路上有多少人骑自行车信息。在本节的项目实例中,将读取上述 CSV 文件中的数据,并实现数据可视化分析功能。本项目的功能模块如图 9-1 所示。

扫码观看本节视频讲解

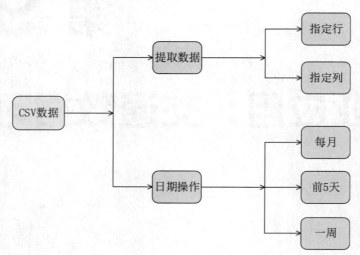

图 9-1 系统模块架构

9.2 从 CSV 文件读取数据

在本节的内容中,将提取 CSV 文件 bikes.csv 中的骑行数据,了解本街道的骑行交通状况。

9.2.1 读取显示 CSV 文件中的前 3 条骑行数据

扫码观看本节视频讲解

在下面的实例文件 002.py 中,读取并显示了文件 bikes.csv 中的前 3 条数据。

源码路径:daima\9\9-2\002.py

```
import pandas as pd
broken_df = pd.read_csv('bikes.csv')
print(broken_df[:3])
```

执行后会输出:

```
    Date;Berri 1;Brébeuf (données non
disponibles);Côte-Sainte-Catherine;Maisonneuve 1;Maisonneuve 2;du
Parc;Pierre-Dupuy;Rachel1;St-Urbain (données non disponibles)
0              01/01/2012;35;;0;38;51;26;10;16;
1              02/01/2012;83;;1;68;153;53;6;43;
2              03/01/2012;135;;2;104;248;89;3;58;
```

读者会发现上述执行效果显得比较凌乱,此时可以利用 read_csv( )方法中的参数选项进行设置。方法 read_csv( )的语法格式如下:

```
pandas.read_csv(filepath_or_buffer, sep=',', delimiter=None, header='infer',
names=None, index_col=None, usecols=None, squeeze=False, prefix=None,
mangle_dupe_cols=True, dtype=None, engine=None, converters=None,
true_values=None, false_values=None, skipinitialspace=False, skiprows=None,
nrows=None, na_values=None, keep_default_na=True, na_filter=True,
verbose=False, skip_blank_lines=True, parse_dates=False,
infer_datetime_format=False, keep_date_col=False, date_parser=None,
dayfirst=False, iterator=False, chunksize=None, compression='infer',
thousands=None, decimal='.', lineterminator=None, quotechar='"', quoting=0,
escapechar=None, comment=None, encoding=None, dialect=None,
tupleize_cols=False, error_bad_lines=True, warn_bad_lines=True, skipfooter=0,
skip_footer=0, doublequote=True, delim_whitespace=False, as_recarray=False,
compact_ints=False, use_unsigned=False, low_memory=True, buffer_lines=None,
memory_map=False, float_precision=None) [source]
```

各个参数的具体说明如下。

- filepath_or_buffer:其值可以是 str、pathlib.Path、py._path.local.LocalPath 或任何具有 read( )方法的对象(如文件句柄或 StringIO)。字符串可以是 URL,有效的 URL 方案包括 http、ftp、s3 和 file。其值文件 URL 需要主机,如本地文件可以是 file://localhost/path/to/table.csv。
- sep: 表示分隔符,可以是 str,默认为逗号','。如果长度大于 1 个字符且与'\s+'不同的分隔符将被解释为正则表达式,将强制使用 Python 解析引擎,并忽略数据中的引号。正则表达式示例:'\r\t'。
- delimiter:str,默认为 None,表示 sep 的备用参数名称。
- delim_whitespace:可以是 boolean 或 default False,用于指定是否将空白(如' '或' ')用作 sep,相当于设置 sep='\s+'。如果将此选项设置为 True,则不应该为 delimiter 参数传入任何内容。
- header:可以是 int 或 ints 列表,默认为'infer',表示用作列名称的行号以及数据的开始。如果未传递 names,默认行就好像设置为 0;否则 None。显式传递 header=0,以便能够替换现有名称。头部可以是整数列表,其指定列上的多索引的行位置[0,1,3]。未指定的插入行将被跳过(如在此示例中跳过 2)。请注意,如果 skip_blank_lines=True,此参数将忽略已注释的行和空行,因此 header = 0 表示数据的第一行,而不是文件的第一行。
- names:可以是 array-like 或 default,表示要使用的列名称列表。如果文件不包含标题行,则应明确传递 header = None。除非 mangle_dupe_cols = True(这是默认值);否则不允许在此列表中重复。

- ► index_col：可以是 int 或序列或 False，默认值无。用作 DataFrame 的行标签的列。如果给出序列，则使用 MultiIndex。如果在每行结尾处都有带分隔符的格式不正确的文件，则可以考虑使用 index_col = False 强制 pandas _not_ 使用第一列作为索引(行名称)。
- ► usecols：可以是 array-like，默认值无。用于返回列的子集。此数组中的所有元素必须是位置(即文档列中的整数索引)或对应于用户在名称中提供或从文档标题行推断的列名称的字符串。例如，有效的 usecols 参数将是[0, 1, 2]或['foo', 'bar', 'baz']。使用此参数会导致更快的解析时间和更低的内存使用率。
- ► as_recarray：可以是 boolean，默认值为 False。
- ► DEPRECATED：此参数将在以后的版本中删除，需要用 pd.read_csv(...) 和 to_records( )代替。在解析数据后，返回 NumPy recarray 而不是 DataFrame。如果设置为 True，此选项优先于 squeeze 参数。此外，由于行索引在此类格式中不可用，因此将忽略 index_col 参数。
- ► squeeze：可以是 boolean，默认值为 False。如果解析的数据只包含一列，则返回一个 Series。
- ► prefix：可以是 str，默认值无。表示在没有标题时添加到列号的前缀，如'X'代表 X0、X1、…。
- ► mangle_dupe_cols：可以是 boolean，默认值为 True。用于重复的列将指定为"X.0"…"X.N"，而不是"X"…"X"。如果在列中存在重复的名称，则传入 False 将导致覆盖数据。
- ► dtype：表示输入列的名称或字典 ->类型，默认值无。
- ► engine：可以是{'c', 'python'}，可选参数，供解析器引擎使用。C 引擎速度更快，而 python 引擎目前更加完善。
- ► converters：可以是 dict，默认值无。说明转换某些列中的值的函数，键可以是整数或列标签。
- ► skipinitialspace：可以是 boolean，默认值为 False，用于跳过分隔符后的空格。
- ► skiprows：可以是 list-like 或 integer，默认值无，表示要跳过的行号(0 索引)或要跳过的行数(int)在文件的开头。
- ► skipfooter：可以是 int，默认值 0，表示跳过文件底部的行数(不支持 engine ='c')。
- ► nrows：可以是 int，默认值无。表示要读取的文件的行数，适用于读取大文件的片段。
- ► na_values：可以是 scalar、str、list-like 或 dict，默认值无。表示可识别为 NA / NaN 的其他字符串。如果 dict 通过，特定的每列 NA 值。在默认情况下，以下值被解释为 NaN：'', '#N/A', '#N/A N/A', '#NA', '-1.#IND', '-1.#QNAN' '-NaN', '-nan', '.#IND', '1.#QNAN', 'N/A', 'NA', 'NULL', 'NaN', 'nan'"。
- ► keep_default_na：可以是 boolean，默认值为 True。如果指定了 na_values 并且 keep_default_na 为 False，则将覆盖默认 NaN 值；否则将追加它们。

- na_filter：可以是 boolean，默认值为 True。用于检测缺失值标记(空字符串和 na_values 的值)。在没有任何 NA 的数据中，传递 na_filter = False 可以提高读取大文件的性能。
- verbose：可以是 boolean，默认值为 False，用于指示放置在非数字列中的 NA 值的数量。
- skip_blank_lines：可以是 boolean，默认值为 True。如果为 True，请跳过空白行，而不是解释为 NaN 值。
- parse_dates：可以是 boolean 或列表或名称或列表或 dict 列表，默认值为 False。如果为 True，则尝试解析索引。
- infer_datetime_format：可以是 boolean，默认值为 False。如果启用了 True 和 parse_dates，pandas 将尝试推断列中 datetime 字符串的格式，如果可以推断，则可以切换到更快的解析方式。在某些情况下，这可以将解析速度提高 5～10 倍。
- keep_date_col：可以是 boolean，默认值为 False。如果 True 和 parse_dates 指定合并多个列，则保留原始列。
- date_parser：一个函数，用于将字符串列序列转换为 datetime 实例数组的函数。默认使用 dateutil.parser.parser 进行转换。Pandas 将尝试以 3 种不同的方式调用 date_parser，如果发生异常，则推进到下一个：①将一个或多个数组(由 parse_dates 定义)作为参数传递；②将由 parse_dates 定义的列中的字符串值连接(逐行)到单个数组中并传递；③对于每一行，使用一个或多个字符串(对应于由 parse_dates 定义的列)作为参数调用 date_parser 一次。
- dayfirst：boolean 类型，默认值为 False。用于返回 TextFileReader 对象以进行迭代或使用 get_chunk( )获取块。
- chunksize：int 类型，用于返回 TextFileReader 对象以进行迭代。在 iterator 和 chunksize 中查看 IO 工具文档了解更多信息。
- compression：{'infer', 'gzip', 'bz2', 'zip', 'xz', None}类型，用于将磁盘上的数据即时解压缩。如果"infer"，则使用 gzip、bz2、zip 或 xz，如果 filepath_or_buffer 则是分别以".gz"".bz2"".zip"或"xz"结尾的字符串，否则不进行解压缩。如果使用'zip'，ZIP 文件必须只包含一个要读入的数据文件。设置为无，无解压缩。
- float_precision：string 类型，用于指定 C 引擎应该为浮点值使用哪个转换器。选项为普通转换器的无、高精度转换器的高和往返转换器的 round_trip。
- lineterminator：str(length 1)类型，用于将文件拆分成行的字符。只有 C 解析器有效。
- quotechar：str(length 1)类型，用于表示带引号项目的开始和结束的字符。引号项可以包含分隔符，它将被忽略。
- doublequote：boolean 类型，默认值为 True。当指定 quotechar 且引用不是 QUOTE_NONE 时，指示是否将一个字段中的两个连续的元素解释为单个 quotechar 元素。
- escapechar：str(length 1)类型，引号时用于转义分隔符的单字符字符串为

QUOTE_NONE。
- encoding：str 类型，在读/写时用于 UTF 的编码(如'utf-8')。
- tupleize_cols：boolean 类型，默认值为 False。用于将列上的元组列表保留为原样(默认是将列转换为多索引)。
- error_bad_lines：boolean 类型，默认值为 True。在默认情况下，具有太多字段的行(如具有太多逗号的 csv 行)将引发异常，并且不会返回 DataFrame。如果为 False，那么这些"坏行"将从返回的 DataFrame 中删除(只有 C 解析器有效)。
- warn_bad_lines：boolean 类型，默认值为 True。如果 error_bad_lines 为 False，并且 warn_bad_lines 为 True，则将输出每个"坏行"的警告(只有 C 解析器有效)。
- low_memory：boolean 类型，默认值为 True。在内部以块的方式处理文件，导致解析时内存使用较少，但可能是混合类型推断。要确保没有混合类型，请设置 False，或使用 dtype 参数指定类型。请注意，无论如何，整个文件都读入单个 DataFrame，请使用 chunksize 或迭代器参数以块形式返回数据(只有 C 解析器有效)。
- compact_ints：boolean 类型，默认值为 False。如果 compact_ints 为 True，则对于任何整数为 dtype 的列，解析器将尝试将其转换为可能的最小整数 dtype，根据 use_unsigned 参数的规范，可以是有符号或无符号。
- use_unsigned：boolean 类型，默认值为 False。如果整数列被压缩(即 compact_ints = True)，请指定该列是否应压缩到最小有符号或无符号整数 dtype。
- memory_map：boolean 类型，默认值为 False。如果为 filepath_or_buffer 提供了文件路径，则将文件对象直接映射到内存上，并从中直接访问数据。使用此选项可以提高性能，因为不再有任何 I/O 开销。

在下面的实例文件 003.py 中，使用规整的格式读取并显示文件 bikes.csv 中的前 3 条数据。

源码路径：daima\9\9-2\003.py

```
import pandas as pd
fixed_df = pd.read_csv('bikes.csv', sep=';', encoding='latin1',
parse_dates=['Date'], dayfirst=True, index_col='Date')
print(fixed_df[:3])
```

执行后会输出：

```
            Berri 1  Brébeuf (données non disponibles)  \
Date
2010-01-01       35                                NaN
2010-01-02       83                                NaN
2010-01-03      135                                NaN

            Côte-Sainte-Catherine  Maisonneuve 1  Maisonneuve 2  du Parc  \
Date
2010-01-01                      0             38             51       26
2010-01-02                      1             68            153       53
2010-01-03                      2            104            248       89
```

```
                Pierre-Dupuy   Rachel1   St-Urbain  (donnÃ©es non disponibles)
Date
2010-01-01            10         16                         NaN
2010-01-02             6         43                         NaN
2010-01-03             3         58                         NaN
```

9.2.2 读取显示 CSV 文件中指定列的数据

在读取 CSV 文件时，得到的是一种由行和列组成的数据帧，可以列出在帧中相同方式的元素。例如，在下面的实例文件 004.py 中，读取并显示了文件 bikes.csv 中的 Berri 1 列的数据。

源码路径：daima\9\9-2\004.py

```
import pandas as pd
fixed_df = pd.read_csv('bikes.csv', sep=';', encoding='latin1',
parse_dates=['Date'], dayfirst=True, index_col='Date')
print(fixed_df['Berri 1'])
```

执行后会输出：

```
Date
2010-01-01      35
2010-01-02      83
2010-01-03     135
……省略部分行数
2010-10-23    4177
2010-10-24    3744
2010-10-25    3735
2010-10-26    4290
2010-10-27    1857
2010-10-28    1310
2010-10-29    2919
2010-10-30    2887
2010-10-31    2634
2010-9-01     2405
2010-9-02     1582
2010-9-03      844
2010-9-04      966
2010-9-05     2247
Name: Berri 1, Length: 310, dtype: int64
```

9.2.3 用统计图可视化 CSV 文件中的数据

为了使应用程序更加美观，在下面的实例文件 005.py 中加入了 matplotlib 功能，以统计图表的方式展示了文件 bikes.csv 中的 Berri 1 列的数据。

源码路径：daima\9\9-2\005.py

```
import pandas as pd
import matplotlib.pyplot as plt
plt.rcParams['figure.figsize'] = (15, 5)
```

```python
fixed_df = pd.read_csv('bikes.csv', sep=';', encoding='latin1',
parse_dates=['Date'], dayfirst=True, index_col='Date')
fixed_df['Berri 1'].plot()
plt.show()
```

执行后会显示每个月的骑行数据统计图，执行效果如图9-2所示。

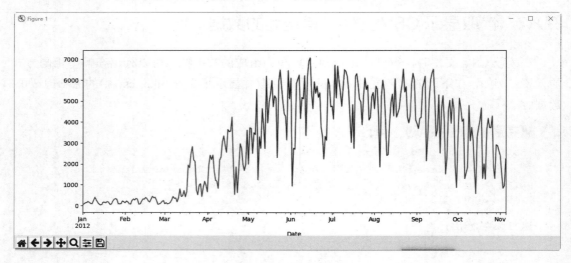

图9-2　执行效果

9.2.4　选择指定数据

请看下面的实例文件006.py，功能是处理一个更大的数据集文件39-service-requests.csv，打印输出这个文件中的数据信息。文件39-service-requests.csv 是 311 的服务请求从纽约开放数据的一个子集，完整文件有 52MB，书中只是截取了一小部分，完整文件可以从网络中获取。

源码路径：daima\9\9-2\006.py

```
import pandas as pd
complaints = pd.read_csv('39-service-requests.csv')
print(complaints)
```

执行后会显示读取文件 39-service-requests.csv 后的结果，并在最后统计数据数目。执行后会输出：

```
      Unique Key            Created Date               Closed Date Agency
Agency Name           Complaint Type
Descriptor                    Location Type  Incident Zip        Incident Address
Street Name     Cross Street 1             Cross Street 2 Intersection Street
1 Intersection Street 2  Address Type              City  Landmark Facility Type
Status              Due Date Resolution Action Updated Date        Community Board
Borough  X Coordinate (State Plane)  Y Coordinate (State Plane)  Park Facility
Name   Park Borough         School Name School Number School Region  School Code
School Phone Number                         School Address  School City
School State   School Zip School Not Found  School or Citywide Complaint  Vehicle
```

```
      Type  Taxi Company  Borough  Taxi Pick Up Location  Bridge Highway Name  Bridge
Highway Direction  Road Ramp  Bridge Highway Segment  Garage Lot Name  Ferry
Direction  Ferry Terminal Name   Latitude   Longitude
Location
0     26589651  10/31/2013 02:08:41 AM               NaN    NYPD           New
York City Police Department   Noise - Street/Sidewalk
Loud Talking              Street/Sidewalk        11432.0          90-03 169 STREET
169 STREET        90 AVENUE                 91 AVENUE              NaN
NaN       ADDRESS         JAMAICA       NaN     Precinct   Assigned
10/31/2013 10:08:41 AM      10/31/2013 02:35:17 AM       12  QUEENS
QUEENS            1042027.0                  197389.0       Unspecified
QUEENS     Unspecified    Unspecified    Unspecified  Unspecified
Unspecified                             Unspecified   Unspecified
Unspecified   Unspecified           N              NaN           NaN
NaN            NaN              NaN            NaN          NaN
NaN          NaN           NaN        NaN  40.708275  -73.791604
(40.70827532593202, -73.791603395779721)
1     26593698  10/31/2013 02:01:04 AM               NaN    NYPD           New
York City Police Department   Illegal Parking
Commercial Overnight Parking          Street/Sidewalk      11378.0
58 AVENUE         58 AVENUE        58 PLACE               59 STREET
NaN         NaN    BLOCKFACE       MASPETH     NaN     Precinct
Open  10/31/2013 10:01:04 AM              NaN        05  QUEENS
QUEENS            1009349.0                  201984.0       Unspecified
QUEENS     Unspecified    Unspecified    Unspecified  Unspecified
Unspecified                             Unspecified   Unspecified
Unspecified   Unspecified           N              NaN           NaN
NaN            NaN              NaN            NaN          NaN
NaN          NaN           NaN        NaN  40.721041  -73.909453
(40.721040535628305, -73.90945306791765)
2     26594139  10/31/2013 02:00:24 AM  10/31/2013 02:40:32 AM   NYPD
New York City Police Department    Noise - Commercial
Loud Music/Party          Club/Bar/Restaurant      10032.0          4060
BROADWAY          BROADWAY   WEST 171 STREET            WEST 172 STREET
NaN         NaN    ADDRESS       NEW YORK     NaN     Precinct
Closed 10/31/2013 10:00:24 AM      10/31/2013 02:39:42 AM    12
MANHATTAN     MANHATTAN           1001088.0                  246531.0
Unspecified    MANHATTAN     Unspecified    Unspecified   Unspecified
Unspecified    Unspecified                             Unspecified
Unspecified   Unspecified   Unspecified         N                   NaN
NaN            NaN              NaN            NaN
NaN         NaN           NaN           NaN          NaN
NaN  40.843330  -73.939144  (40.84332975466513, -73.93914371913482)
3     26595721  10/31/2013 01:56:23 AM  10/31/2013 02:21:48 AM   NYPD
New York City Police Department    Noise - Vehicle
Car/Truck Horn            Street/Sidewalk       10023.0          WEST 72
STREET      WEST 72 STREET   COLUMBUS AVENUE        AMSTERDAM AVENUE
NaN         NaN    BLOCKFACE       NEW YORK    NaN     Precinct
Closed 10/31/2013 09:56:23 AM      10/31/2013 02:21:10 AM    07
MANHATTAN     MANHATTAN            989730.0                  222727.0
Unspecified    MANHATTAN     Unspecified    Unspecified   Unspecified
Unspecified    Unspecified                             Unspecified
Unspecified   Unspecified   Unspecified         N           NaN
NaN            NaN              NaN            NaN
NaN          NaN           NaN
```

```
NaN          NaN                    NaN           NaN           NaN
NaN 40.778009 -73.980213         (40.7780087446372, -73.98021349023975)
......省略部分执行结果
[263 rows x 52 columns]
```

而在下面的实例文件 007.py 中，首先输出显示了文件 39-service-requests.csv 中 Complaint Type 列的信息，然后输出了文件 39-service-requests.csv 中的前 5 行信息，其次输出了文件 39-service-requests.csv 中前 5 行 Complaint Type 列的信息，再次又输出了文件 39-service-requests.csv 中 Complaint Type 和 Borough 这两列的信息，最后输出了文件 39-service-requests.csv 中 Complaint Type 和 Borough 这两列的前 10 行信息。

源码路径：**daima\9\9-2\007.py**

```
import pandas as pd
complaints = pd.read_csv('39-service-requests.csv')
print(complaints['Complaint Type'])
print(complaints[:5])
print(complaints[:5]['Complaint Type'])
print(complaints[['Complaint Type', 'Borough']])
print(complaints[['Complaint Type', 'Borough']][:10])
```

执行后会输出：

```
//下面首先输出 Complaint Type 列的信息
0          Noise - Street/Sidewalk
1                  Illegal Parking
2                Noise - Commercial
3                   Noise - Vehicle
4                            Rodent
5                Noise - Commercial
6                  Blocked Driveway
7                Noise - Commercial
8                Noise - Commercial
9                Noise - Commercial
10         Noise - House of Worship
11               Noise - Commercial
12                 Illegal Parking
13                  Noise - Vehicle
14                           Rodent
15         Noise - House of Worship
16         Noise - Street/Sidewalk
17                 Illegal Parking
18            Street Light Condition
19               Noise - Commercial
20         Noise - House of Worship
21               Noise - Commercial
22                  Noise - Vehicle
23               Noise - Commercial
24                 Blocked Driveway
25         Noise - Street/Sidewalk
26            Street Light Condition
27             Harboring Bees/Wasps
28         Noise - Street/Sidewalk
29            Street Light Condition
```

```
                    ...
233           Noise - Commercial
234              Taxi Complaint
235          Sanitation Condition
236       Noise - Street/Sidewalk
237           Consumer Complaint
238       Traffic Signal Condition
239         DOF Literature Request
240         Litter Basket / Request
241             Blocked Driveway
242       Violation of Park Rules
243        Collection Truck Noise
244              Taxi Complaint
245              Taxi Complaint
246         DOF Literature Request
247       Noise - Street/Sidewalk
248              Illegal Parking
249              Illegal Parking
250             Blocked Driveway
251      Maintenance or Facility
252           Noise - Commercial
253              Illegal Parking
254                       Noise
255                      Rodent
256              Illegal Parking
257                       Noise
258       Street Light Condition
259                Noise - Park
260             Blocked Driveway
261              Illegal Parking
262           Noise - Commercial
Name: Complaint Type, Length: 263, dtype: object
```
//下面输出前5列信息
```
   Unique Key          Created Date              Closed Date Agency  \
0    26589651  10/31/2013 02:08:41 AM                      NaN   NYPD
1    26593698  10/31/2013 02:01:04 AM                      NaN   NYPD
2    26594139  10/31/2013 02:00:24 AM  10/31/2013 02:40:32 AM   NYPD
3    26595721  10/31/2013 01:56:23 AM  10/31/2013 02:21:48 AM   NYPD
4    26590930  10/31/2013 01:53:44 AM                      NaN  DOHMH

                              Agency Name           Complaint Type  \
0             New York City Police Department  Noise - Street/Sidewalk
1             New York City Police Department          Illegal Parking
2             New York City Police Department       Noise - Commercial
3             New York City Police Department           Noise - Vehicle
4   Department of Health and Mental Hygiene                    Rodent

                  Descriptor       Location Type  Incident Zip  \
0                Loud Talking     Street/Sidewalk       11432.0
1  Commercial Overnight Parking  Street/Sidewalk       11378.0
2            Loud Music/Party   Club/Bar/Restaurant      10032.0
3               Car/Truck Horn    Street/Sidewalk       10023.0
4  Condition Attracting Rodents         Vacant Lot       10027.0
```

```
   Incident Address                      ...                                    \
0  90-03 169 STREET                      ...
1         58 AVENUE                      ...
2     4060 BROADWAY                      ...
3    WEST 72 STREET                      ...
4   WEST 124 STREET                      ...

  Bridge Highway Name Bridge Highway Direction Road Ramp  \
0                 NaN                      NaN       NaN
1                 NaN                      NaN       NaN
2                 NaN                      NaN       NaN
3                 NaN                      NaN       NaN
4                 NaN                      NaN       NaN

  Bridge Highway Segment Garage Lot Name Ferry Direction Ferry Terminal Name
\
0                    NaN            NaN             NaN                 NaN
1                    NaN            NaN             NaN                 NaN
2                    NaN            NaN             NaN                 NaN
3                    NaN            NaN             NaN                 NaN
4                    NaN            NaN             NaN                 NaN

    Latitude  Longitude                                     Location
0  40.708275 -73.791604  (40.70827532593202, -73.79160395779721)
1  40.721041 -73.909453  (40.721040535628305, -73.90945306791765)
2  40.843330 -73.939144  (40.84332975466513, -73.93914371913482)
3  40.778009 -73.980213  (40.7780087446372, -73.98021349023975)
4  40.807691 -73.947387  (40.80769092704951, -73.94738703491433)

[5 rows x 52 columns]
```
//下面输出前 5 行 Complaint Type 列的信息
```
[5 rows x 52 columns]
0     Noise - Street/Sidewalk
1             Illegal Parking
2          Noise - Commercial
3             Noise - Vehicle
4                      Rodent
```
······省略部分
```
259            Noise - Park        BROOKLYN
260        Blocked Driveway          QUEENS
261         Illegal Parking        BROOKLYN
262      Noise - Commercial       MANHATTAN
[263 rows x 2 columns]
```
//下面输出 Complaint Type 和 Borough 这两列的信息
```
            Complaint Type    Borough
0  Noise - Street/Sidewalk     QUEENS
1          Illegal Parking     QUEENS
2       Noise - Commercial  MANHATTAN
3          Noise - Vehicle  MANHATTAN
4                   Rodent  MANHATTAN
5       Noise - Commercial     QUEENS
```
//下面输出 Complaint Type 和 Borough 这两列的前 10 行信息

```
         Complaint Type      Borough
0  Noise - Street/Sidewalk    QUEENS
1        Illegal Parking      QUEENS
2       Noise - Commercial  MANHATTAN
3        Noise - Vehicle    MANHATTAN
4                 Rodent    MANHATTAN
5       Noise - Commercial    QUEENS
6         Blocked Driveway    QUEENS
7       Noise - Commercial    QUEENS
8       Noise - Commercial  MANHATTAN
9       Noise - Commercial   BROOKLYN
```

在下面的实例文件 008.py 中，首先输出显示了文件 39-service-requests.csv 中 Complaint Type 列中数值前 10 名的信息，然后在图表中统计显示这前 10 名信息。

源码路径：daima\9\9-2\008.py

```python
import pandas as pd
import matplotlib.pyplot as plt

pd.set_option('display.width', 5000)
pd.set_option('display.max_columns', 60)

plt.rcParams['figure.figsize'] = (10, 6)

complaints = pd.read_csv('39-service-requests.csv')
complaint_counts = complaints['Complaint Type'].value_counts()
print(complaint_counts[:10])#打印输出 Complaint Type 列中数值前 10 名的信息
complaint_counts[:10].plot(kind='bar')#绘制 Complaint Type 列中数值前 10 名的图表信息
plt.show()
```

执行后会在控制台中输出显示 Complaint Type 列中数值前 10 名的信息：

```
Noise - Commercial          51
Noise                       27
Noise - Street/Sidewalk     22
Blocked Driveway            21
Illegal Parking             18
Taxi Complaint              13
Traffic Signal Condition    10
Rodent                      10
Water System                 9
Noise - Vehicle              7
Name: Complaint Type, dtype: int64
```

并且执行后还会在 Matplotlib 图表中统计列 Complaint Type 中数值前 10 名的信息，执行效果如图 9-3 所示。

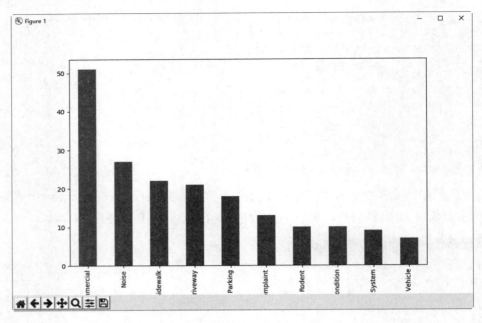

图 9-3 执行效果

9.3 日期相关操作

在进行数据统计分析时，时间通常是一个重要的因素之一。本节将详细讲解和日期相关的操作知识，为读者步入本书后面知识的学习打下基础。

9.3.1 统计每个月的骑行数据

扫码观看本节视频讲解

在下面的实例文件 009.py 中，可以使用 matplotlib 统计出文件 bikes.csv 中每个月的骑行数据信息。

源码路径：codes\9\9-3\009.py

```python
import pandas as pd
import matplotlib.pyplot as plt

plt.rcParams['figure.figsize'] = (10, 8)
plt.rcParams['font.family'] = 'sans-serif'

pd.set_option('display.width', 5000)
pd.set_option('display.max_columns', 60)

bikes = pd.read_csv('bikes.csv', sep=';', encoding='latin1',
parse_dates=['Date'], dayfirst=True, index_col='Date')
bikes['Berri 1'].plot()
plt.show()
```

执行后的效果如图 9-4 所示。

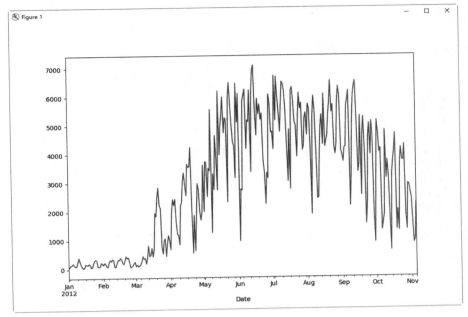

图 9-4　执行后的效果

9.3.2　展示某街道前 5 天的骑行数据信息

在下面的实例文件 010.py 中，首先输出显示文件 bikes.csv 中 Berri 1 街道前 5 天的骑行数据信息；然后使用 print(berri_bikes.index)输出了星期几的时间。

源码路径：codes\9\9-3\010.py

```python
import pandas as pd

bikes = pd.read_csv('bikes.csv', sep=';', encoding='latin1',
parse_dates=['Date'], dayfirst=True, index_col='Date')
berri_bikes = bikes[['Berri 1']].copy()
print(berri_bikes[:5])
print(berri_bikes.index)
```

执行后会输出：

```
            Berri 1
Date
2010-01-01       35
2010-01-02       83
2010-01-03      135
2010-01-04      144
2010-01-05      197
DatetimeIndex(['2010-01-01', '2010-01-02', '2010-01-03', '2010-01-04',
       '2010-01-05', '2010-01-06', '2010-01-07', '2010-01-08',
       '2010-01-09', '2010-01-10',
       ...
```

```
                '2010-10-27', '2010-10-28', '2010-10-29', '2010-10-30',
                '2010-10-31', '2010-9-01', '2010-9-02', '2010-9-03',
                '2010-9-04', '2010-9-05'],
               dtype='datetime64[ns]', name='Date', length=310, freq=None)
```

由上述执行效果可知，只输出显示了 310 天的统计数据。其实 pandas 有一系列非常好的时间序列功能，所以如果想得到每一行的月份，可以通过以下文件 09.py 实现。

源码路径：daima\9\9-3\09.py

```python
import pandas as pd

bikes = pd.read_csv('bikes.csv', sep=';', encoding='latin1',
parse_dates=['Date'], dayfirst=True, index_col='Date')
berri_bikes = bikes[['Berri 1']].copy()
print(berri_bikes.index.day)
print(berri_bikes.index.weekday)
```

执行后会输出：

```
Int64Index([ 1,  2,  3,  4,  5,  6,  7,  8,  9, 10,
            ...
            27, 28, 29, 30, 31,  1,  2,  3,  4,  5],
           dtype='int64', name='Date', length=310)
Int64Index([6, 0, 1, 2, 3, 4, 5, 6, 0, 1,
            ...
            5, 6, 0, 1, 2, 3, 4, 5, 6, 0],
           dtype='int64', name='Date', length=310)
```

在上述输出结果中，0 表示星期一。可以使用 pandas 灵活获取某一天是星期几，请看下面的实例文件 012.py。

源码路径：daima\9\9-3\012.py

```python
import pandas as pd

bikes = pd.read_csv('bikes.csv', sep=';', encoding='latin1',
parse_dates=['Date'], dayfirst=True, index_col='Date')
berri_bikes = bikes[['Berri 1']].copy()
berri_bikes.loc[:,'weekday'] = berri_bikes.index.weekday
print(berri_bikes[:5])
```

执行后会输出：

```
            Berri 1  weekday
Date
2010-01-01       35        6
2010-01-02       83        0
2010-01-03      135        1
2010-01-04      144        2
2010-01-05      197        3
```

9.3.3 统计周一到周日每天的数据

在现实应用中，也可以统计周一到周日每天的统计数据。例如，在下面的实例文件

013.py 中,首先显示了周一到周日每天的统计数据,然后用更加通俗易懂的星期几的英文名显示了周一到周日每天的骑行统计数据。

源码路径:daima\9\9-3\013.py

```python
import pandas as pd
bikes = pd.read_csv('bikes.csv', sep=';', encoding='latin1',
parse_dates=['Date'], dayfirst=True, index_col='Date')
berri_bikes = bikes[['Berri 1']].copy()
berri_bikes.loc[:,'weekday'] = berri_bikes.index.weekday

weekday_counts = berri_bikes.groupby('weekday').aggregate(sum)
print(weekday_counts)

weekday_counts.index = ['Monday', 'Tuesday', 'Wednesday', 'Thursday', 'Friday',
'Saturday', 'Sunday']
print(weekday_counts)
```

执行后会输出:

```
         Berri 1
weekday
0        134298
1        135305
2        152972
3        160131
4        141771
5        101578
6         99310
          Berri 1
Monday    134298
Tuesday   135305
Wednesday 152972
Thursday  160131
Friday    141771
Saturday  101578
Sunday     99310
```

9.3.4 使用 matplotlib 图表可视化展示统计数据

为了使统计数据更加直观,可以在程序中使用 matplotlib 技术。例如,在下面的实例文件 014.py 中,使用 matplotlib 图表统计了周一到周日每天的骑行数据。

源码路径:daima\9\9-3\014.py

```python
import pandas as pd
import matplotlib.pyplot as plt
plt.rcParams['figure.figsize'] = (15, 5)
bikes = pd.read_csv('bikes.csv',
                    sep=';', encoding='latin1',
                    parse_dates=['Date'], dayfirst=True,
                    index_col='Date')
# 添加标识
berri_bikes = bikes[['Berri 1']].copy()
```

```
berri_bikes.loc[:,'weekday'] = berri_bikes.index.weekday

# 开始统计
weekday_counts = berri_bikes.groupby('weekday').aggregate(sum)
weekday_counts.index = ['Monday', 'Tuesday', 'Wednesday', 'Thursday', 'Friday',
'Saturday', 'Sunday']
weekday_counts.plot(kind='bar')

plt.show()
```

执行后的效果如图9-5所示。

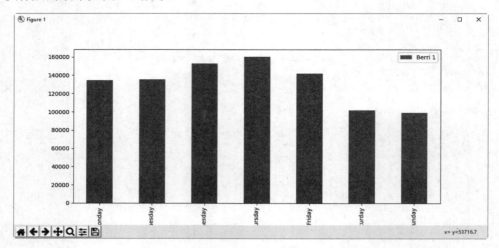

图9-5 执行后的效果

再看下面的实例文件015.py，借助素材文件weather_2012.csv，使用matplotlib统计了加拿大2012年的全年天气数据信息。

源码路径：codes\9\9-3\015.py

```
import pandas as pd
import matplotlib.pyplot as plt
import numpy as np

plt.rcParams['figure.figsize'] = (15, 3)
plt.rcParams['font.family'] = 'sans-serif'
weather_2012_final = pd.read_csv('weather_2012.csv', index_col='Date/Time')
weather_2012_final['Temp (C)'].plot(figsize=(15, 6))
plt.show()
```

执行后的效果如图9-6所示。

第 9 章 商业应用：交通数据可视化

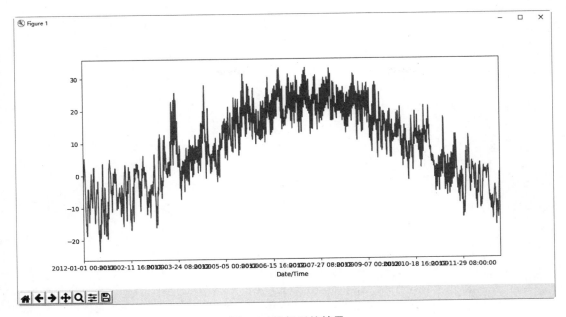

图 9-6 执行后的效果

第 10 章

商业应用：招聘信息可视化

  21 世纪什么最重要？人才最重要。在现实世界中，各种企事业用人单位为了获得更好的发展，纷纷通过各种途径招纳人才。在众多招贤纳士的途径中，招聘网成为最重要的渠道之一。无论对于用人单位还是应聘者，数据分析招聘网的招聘信息是十分重要的。本章将通过一个综合实例的实现过程，详细讲解爬虫抓取某知名招聘网中招聘信息的方法，并讲解可视化分析招聘信息的过程。

10.1 系统背景介绍

在当今社会环境下，招聘已然成为人力资源管理的热点，猎头公司、招聘网站、人才测评等配套服务机构应运而生，其核心在于为企业提供了人才信息渠道。这些专业机构为企业提供专业服务，猎头公司、招聘网站解决的是"符合企业要求的人才在哪里"的问题，人才测评公司解决的是"这个人到底有何素质、适合做什么"的问题。随着企业用人需求的弹性化和动态变化，需要企业内部专业经理们与外部专业机构解决"企业到底需要什么样的人"这一问题。

扫码观看本节视频讲解

随着时代的发展，很多公司在招聘时都会收到成千上万的简历，如何挑选合适的应聘者成为公司比较棘手的事情，这给招聘单位的人事部门带来相当大的工作负担。与其他传统的人才中介相比，网上招聘具有成本低、容量大、速度快和强调个性化服务的优势。伴随着新增岗位源源不断涌现，即使广泛存在的岗位对人的要求也变得模糊起来。对于用人单位的人力资源部来说，及时了解招聘行情是自己最基本的业务范畴。而对于应聘者来说，根据自己的情况选择合适待遇的用人单位是自己的首要应聘目的。此时，将招聘网中的招聘信息进行可视化处理就变得十分重要了。下面以某开发公司招聘 Python 开发工程师为例进行说明。

- 开发公司的人力资源部可以可视化分析招聘网中的与 Python 相关的招聘信息，了解不同学历和不同工作经验对应的薪资水平。
- Python 应聘者通过可视化分析招聘网中的与 Python 相关的招聘信息，了解不同学历和不同工作经验对应的薪资水平。

目前可视化招聘信息已经成为人力资源经理关注的焦点，在特定的发展阶段、特定的文化背景下，面对变动的市场环境和弹性的岗位要求，企业到底需要什么样的人，为不同层次的人才提供什么样的待遇是他们格外关注的问题。基于目前招聘信息可视化需求分析的重要性，很多专业招聘网和猎头机构热衷于用可视化图表展示人才供求状况，可视化分析招聘信息大有可为。

10.2 系统架构分析

本项目首先用爬虫抓取某知名招聘网中招聘信息，然后可视化分析抓取到的招聘信息，为用人单位人力资源部和应聘者提供强有力的数据支撑。本项目的功能模块如图 10-1 所示。

在图 10-1 中各个模块的具体说明如下。

1) 系统设置

设置使用 MySQL 数据库保存用爬虫抓取到的数据，然后用

扫码观看本节视频讲解

Flask Web 框架提取数据库中的数据，用网页的形式可视化展示招聘信息。

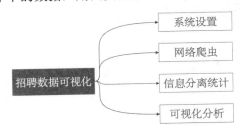

图 10-1　系统模块架构

2）网络爬虫

根据用户输入的关键字，使用网络爬虫技术抓取招聘网中的招聘信息，将抓取到的招聘信息添加到 MySQL 数据库中。

3）信息分离统计

提取在 MySQL 数据库中保存的爬虫数据，分别根据"工作地区""工作经验""薪资水平"和"学历水平"提取并分离招聘信息。

4）可视化分析

提取在 MySQL 数据库中保存的爬虫数据，然后使用开源框架 Highcharts 绘制柱状图和饼状图，可视化展示招聘数据信息。

10.3　系统设置

本项目使用 MySQL 数据库保存用爬虫抓取到的数据，编写程序文件 config.py，设置连接 MySQL 数据库的参数，具体实现代码如下：

扫码观看本节视频讲解

```
HOST = '127.0.0.1'
PORT = '3306'
USERNAME = 'root'
PASSWORD = '66688888'
DATABASE = 'u1'
DB_URI = 
'mysql+pymysql://{username}:{password}@{host}:{port}/{db}?charset=utf8mb4'.
format(username=USERNAME, password=PASSWORD, host=HOST, port=PORT, 
db=DATABASE)
```

10.4　网络爬虫

在本项目网络爬虫模块的实现文件是 data.py，功能是根据输入的关键字抓取招聘网中的招聘信息，将抓取到的招聘信息添加到 MySQL 数据库中。

扫码观看本节视频讲解

10.4.1 建立和数据库的连接

因为需要将爬取的数据添加到 MySQL 数据库中，所以需要导入在前面配置文件 config.py 中设置的数据库连接参数。对应的实现代码如下：

```
host = config.HOST
post = config.PORT
username = config.USERNAME
password = config.PASSWORD
database = config.DATABASE

etree = html.etree
tlock=threading.Lock()


# 拿到游标
cursor = db.cursor()
```

10.4.2 设置 HTTP 请求头 User-Agent

User-Agent 会告诉网站服务器，访问者是通过什么工具来请求的，如果是爬虫请求，一般会拒绝，如果是用户浏览器就会应答。通过设置 HTTP 请求头 User-Agent，可以确保爬取到数据。对应的实现代码如下：

```
user_agent = [

  # Firefox
  "Mozilla/5.0 (Windows NT 6.1; WOW64; rv:34.0) Gecko/20100101 Firefox/34.0",
  "Mozilla/5.0 (X11; U; Linux x86_64; zh-CN; rv:1.10.2.10) Gecko/20100922 Ubuntu/10.10 (maverick) Firefox/3.6.10",
  # Safari
  "Mozilla/5.0 (Windows NT 6.1; WOW64) AppleWebKit/534.57.2 (KHTML, like Gecko) Version/5.1.7 Safari/534.57.2",
  # chrome
  "Mozilla/5.0 (Windows NT 6.1; WOW64) AppleWebKit/537.36 (KHTML, like Gecko) Chrome/310.0.2171.71 Safari/537.36",
  "Mozilla/5.0 (X11; Linux x86_64) AppleWebKit/537.11 (KHTML, like Gecko) Chrome/23.0.1271.64 Safari/537.11",
  "Mozilla/5.0 (Windows; U; Windows NT 6.1; en-US) AppleWebKit/534.16 (KHTML, like Gecko) Chrome/10.0.648.133 Safari/534.16",
  # 360
  "Mozilla/5.0 (Windows NT 6.1; WOW64) AppleWebKit/537.36 (KHTML, like Gecko) Chrome/30.0.15910.101 Safari/537.36",
  "Mozilla/5.0 (Windows NT 6.1; WOW64; Trident/7.0; rv:11.0) like Gecko",
  # 猎豹浏览器
  "Mozilla/5.0 (Windows NT 6.1; WOW64) AppleWebKit/537.1 (KHTML, like Gecko) Chrome/21.0.1180.71 Safari/537.1 LBBROWSER",
  "Mozilla/5.0 (compatible; MSIE 10.0; Windows NT 6.1; WOW64; Trident/5.0; SLCC2; .NET CLR 2.0.50727; .NET CLR 3.5.30729; .NET CLR 3.0.30729; Media Center PC 6.0; .NET4.0C; .NET4.0E; LBBROWSER)",
```

```
    "Mozilla/4.0 (compatible; MSIE 6.0; Windows NT 5.1; SV1; QQDownload
732; .NET4.0C; .NET4.0E; LBBROWSER)",
    # QQ 浏览器
    "Mozilla/5.0 (compatible; MSIE 10.0; Windows NT 6.1; WOW64; Trident/5.0;
SLCC2; .NET CLR 2.0.50727; .NET CLR 3.5.30729; .NET CLR 3.0.30729; Media Center
PC 6.0; .NET4.0C; .NET4.0E; QQBrowser/7.0.3698.400)",
    "Mozilla/4.0 (compatible; MSIE 6.0; Windows NT 5.1; SV1; QQDownload
732; .NET4.0C; .NET4.0E)"
]
def get_user_agent():
    """随机获取一个请求头"""
    return {'User-Agent': random.choice(user_agent)}


def requst(url):
    """requests 到 url 的 HTML"""
    html = requests.get(url,headers=get_user_agent())
    html.encoding = 'gbk'
    return etree.HTML(html.text)
```

10.4.3 抓取信息

分别编写函数 def get_url(url)和 def get_data(urls)，根据设置的要抓取的 URL 地址，分别抓取目标 URL 中的每个招聘信息中的职位名称、公司、城市地区、经验、发布日期、学历、薪资、招聘详情等信息。对应的实现代码如下：

```
def get_url(url):
    data = requst(url)
    href = data.xpath('//*[@id="resultList"]/div/p/span/a/@href')
    print(len(href))
    return href

def get_data(urls):
    """获取 xxjob 职位信息,并存入数据库"""
    list_all = []
    for url in urls:
        regjob = re.compile(r'https://(.*?)51job.com', re.S)
        it = re.findall(regjob, url)
        if it != ['jobs.']:
            print('不匹配')
            continue
        try:
            data = requst(url)
            # 职位名称
            titles = data.xpath('/html/body/div[3]/div[2]/div[2]/div/div[1]/
h1/@title')[0]
            # 公司
            company = data.xpath('/html/body/div[3]/div[2]/div[2]/div/
div[1]/p[1]/a[1]/@title')[0]
```

```python
        ltype = data.xpath('/html/body/div[3]/div[2]/div[2]/
div/div[1]/p[2]/@title')[0]
        ltype_str = "".join(ltype.split())
        # print(ltype_str)
        ltype_list = ltype_str.split('|')
        #城市地区
        addres = ltype_list[0]
        #经验
        exper = ltype_list[1]
        # 发布日期
        if len(ltype_list)>=5:
            #学历
            edu = ltype_list[2]
            dateT = ltype_list[4]
        else:
            #学历
            edu = "没有要求"
            dateT = ltype_list[-1]
        # 薪资
        salary = data.xpath('/html/body/div[3]/div[2]/div[2]/
div/div[1]/strong/text()')
        if len(salary) == 0:
            salary_list = [0,0]
        else:
            salary_list = salary_alter(salary)[0]
        # 招聘详情
        contents = data.xpath('/html/body/div[3]/div[2]/div[3]/
div[1]/div')[0]
        content = contents.xpath('string(.)')
        # content = content.replace(' ','')
        content = "".join(content.split())

        list_all.append([titles,company,addres,salary_list[0],
salary_list[1],dateT,edu,exper,content])
        item = [titles, company, addres, salary_list[0], salary_list[1],
dateT, edu, exper, content]
        write_db(item)
    except:
        print('爬取失败')
```

10.4.4 将抓取的信息添加到数据库

编写函数 write_db(data)，功能是将抓取到的招聘信息添加到 MySQL 数据库中。对应的实现代码如下：

```python
def write_db(data):
    """写入数据库"""
    print(data)
    try:
        tlock.acquire()
        # rows 变量得到数据库中被影响的数据行数
        rescoun = cursor.execute(
```

```
            "insert into data (post,company,address,salary_min,salary_max,
dateT,edu,exper,content) values(%s,%s,%s,%s,%s,%s,%s,%s,%s)", data)
        # 向数据库提交
        db.commit()
        tlock.release()
        # 如果没有commit(),库中字段已经向下移位但内容没有写进,可是自动生成的ID会自动增加
        print('成功')
        global db_item
        db_item = db_item + 1

    except:
        # 发生错误时回滚
        db.rollback()
        tlock.release()
        print('插入失败')
```

10.4.5 处理薪资数据

为了便于在可视化图表中展示薪资水平,向数据库中添加的是单位为元的数据,在图表中展示的薪资单位是千元,如果招聘信息没有工资数据则向数据库添加0。如果在招聘信息中提供的是年薪,则向数据库中添加除以12的月薪。对应的实现代码如下:

```
def salary_alter(salarys):    #[]
    salary_list = []
    for salary in salarys:
        # print(salary)
        if salary == '':
            a = [0,0]
        re_salary = re.findall('[\d+\.\d]*', salary)  # 提取数值--是文本值
        salary_min = float(re_salary[0])  # 将文本转化成数值型,带有小数,用float()

        wan = lambda x,y : [x*10000,y*10000]
        qian = lambda x, y: [x * 1000, y * 1000]
        wanqian = lambda x,y,s :wan(x,y) if '万' in salary else qian(x,y)
        tian = lambda x, y, s: [x,y] if '元' in salary else qian(x, y)

        if '年' in salary:
            salary_max = float(re_salary[2])
            a = wanqian(salary_min,salary_max,salary)
            a[0] = round(a[0] / 12, 2)
            a[1] = round(a[1] / 12, 2)
        elif '月' in salary:
            salary_max = float(re_salary[2])
            a = wanqian(salary_min, salary_max, salary)
        elif '天' in salary:
            salary_max = float(re_salary[0])
            a = tian(salary_min, salary_max, salary)
            a[0] *= 31
            a[1] *= 31
        salary_list.append(a)
    return salary_list
```

10.4.6 清空数据库数据

为了保证可视化分析的数据具有时效性和准确性，在每一次新的抓取之前清空以前抓取的数据。对应的实现代码如下：

```python
def data_clr():
    # 清空data
    try:
        tlock.acquire()
        query = "truncate table 'data'"
        cursor.execute(query)
        db.commit()
        tlock.release()
        print('data 原表已清空')
    except Exception as aa:
        print(aa)
        print('无data表！')
```

10.4.7 执行爬虫程序

设置要爬取的 URL 地址，开始执行爬虫程序。对应的实现代码如下：

```python
def main(kw, city,startpage):
    print(kw)
    print(city)
    city_id = get_cityid(city)
    print(city_id)
    data_clr()
    page = startpage
    global db_item
    db_item = 0
    while(db_item <= 200):
        url = "https://search.51job.com/list/{},000000,0000,00,9,99,{},2,{}.html".format(city_id,kw,page)
        print(url)
        # url = 'https://search.51job.com/jobsearch/search_result.php'
        a =get_url(url)
        print(a)
        get_data(a)
        page = page + 1
        print(db_item)
        # time.sleep(0.2)
```

10.5 信息分离统计

提取在 MySQL 数据库中保存的爬虫数据，分别根据"工作地区""工作经验""薪资水平"和"学历水平"提取招聘信息并进行统计，最终的统计结果将作为后面可视化图表的素材数据。

扫码观看本节视频讲解

10.5.1 根据"工作经验"分析数据

编写程序文件 jinyan.py，根据用人单位的"工作经验"要求提取并统计招聘信息，具体实现代码如下：

```python
def get_edu():
    try:
        cursor.execute("select exper from data ")
        salary = cursor.fetchall()
        # 向数据库提交
        db.commit()
        return salary
    except:
        # 发生错误时回滚
        db.rollback()
        print("查询失败")
        return 0

def jinyanfun():
    edu = get_edu()
    data = []
    for i in edu:
        # print(type(i))
        year = re.findall(r"\d+", i[0])
        if len(year)==1:
            data.append(year[0]+'年工作经验')
        elif len(year)==2:
            data.append(year[0]+'-'+year[1]+'年工作经验')
        elif len(year)==0:
            data.append('无工作经验')

    data = DataFrame(data)
    da = data[0].value_counts()
    # print(da)
    list_all = []
    for (i, j) in zip(da.index, da):
        print(j,i)
        list_all.append([i,j])
    # print(type(da))
    return list_all

if __name__ == '__main__':
    a = jinyanfun()
    print(a)
```

执行后会基于当前抓取到的招聘信息提取"工作经验"列的数据，并进行数据统计。

```
76 3-4 年工作经验
56 2 年工作经验
37 1 年工作经验
```

```
33 无工作经验
14 5-7年工作经验
1 3年工作经验
[['3-4年工作经验', 76], ['2年工作经验', 56], ['1年工作经验', 37], ['无工作经验', 33],
['5-7年工作经验', 14], ['3年工作经验', 1]]
```

10.5.2 根据"工作地区"分析数据

编写程序文件 map.py,根据用人单位的"工作地区"要求提取并统计招聘信息,具体实现代码如下:

```
def get_xinzi():
    try:
        cursor.execute("select address from data ")
        salary = cursor.fetchall()
        # 向数据库提交
        db.commit()
        return salary
    except:
        # 发生错误时回滚
        db.rollback()
        print("查询失败")
        return 0
a = get_xinzi()

list = []
for i in a:
    test = i[0].split('-')
    list.append(test[0])
    # print(test)

data = DataFrame(list)
da = data[0].value_counts()


for (i,j) in zip(da.index,da):
    pass
    print(j,i)
print(type(da))
```

执行后会基于当前抓取到的招聘信息提取"工作地区"列的数据,并进行数据统计。

```
169 广州
14 佛山
12 东莞
8 珠海
4 中山
3 深圳
1 南昌
1 汕头
1 上海
1 惠州
```

```
1 澄迈
1 韶关
1 广东省
```

10.5.3 根据"薪资水平"分析数据

编写程序文件 xinzi.py，根据用人单位的"薪资水平"要求提取并统计招聘信息。在本项目中将工资分为以下 8 个档次：

- 小于 5k；
- 5k~8k；
- 8k~11k；
- 11k~14k；
- 14k-17k；
- 17k-20k；
- 20k-23k；
- 高于 23K。

文件 xinzi.py 的具体实现代码如下：

```python
def get_xinzi():
    try:
        cursor.execute("select salary_min,salary_max from data ")
        salary = cursor.fetchall()
        # 向数据库提交
        db.commit()
        return salary
    except:
        # 发生错误时回滚
        db.rollback()
        print("查询失败")
        return 0

def xinzi():
    a = get_xinzi()
    data = []
    for i in a:
        data.append((int(i[0])+int(i[1]))/2)

    fenzu=pd.cut(data,[0,5000,8000,11000,14000,17000,20000,23000,9000000000],right=False)
    pinshu=fenzu.value_counts()
    # print(pinshu)
    list = []
    for i in pinshu:
        # print(i)
        list.append(i)

    list_all = [
        ['小于5k', list[0]],
```

```
            ['5k~8k', list[1]],
            ['8k~11k',list[2]],
            ['11k~14k',list[3]],
            ['14k-17k', list[4]],
            ['17k-20k', list[5]],
            ['20k-23k', list[6]],
            ['23K~', list[7]]
        ]
        return list

if __name__ == '__main__':
    a = xinzi()
    print(a)
```

执行后会基于当前抓取到的招聘信息提取"薪资水平"列的数据,并进行数据统计。

[11, 34, 47, 53, 31, 14, 15, 12]

10.5.4 根据"学历水平"分析数据

编写程序文件 xueli.py,根据用人单位的"学历水平"要求提取并统计招聘信息,具体实现代码如下:

```
def get_edu():
    try:
        cursor.execute("select edu from data ")
        salary = cursor.fetchall()
        # 向数据库提交
        db.commit()
        return salary
    except:
        # 发生错误时回滚
        db.rollback()
        print("查询失败")
        return 0

def xuelifun():
    edu = get_edu()
    data = []
    for i in edu:
        data.append(i)
        # print(i)

    data = DataFrame(data)
    da = data[0].value_counts()
    # print(da)
    list_all = []
    for (i, j) in zip(da.index, da):
        print(j, i)
        if '招' in i:
            # print(i)
            continue
```

```
        list_all.append([i, j])
    # print(type(da))
    return list_all
if __name__ == '__main__':
    a = xuelifun()
    print(a)
    print(type(a))
```

执行后会基于当前抓取到的招聘信息提取"学历水平"列的数据，并进行数据统计。

```
107 本科
73 大专
26 没有要求
5 硕士
2 中专
2 招若干人
1 招4人
1 招2人
[['本科', 107], ['大专', 73], ['没有要求', 26], ['硕士', 5], ['中专', 2]]
```

10.6 数据可视化

提取在 MySQL 数据库中保存的爬虫数据，并使用在前面分析统计步骤中得到的统计结果，使用开源框架 Highcharts 绘制柱状图和饼状图，使用 Flask 框架可视化展示招聘数据信息。

扫码观看本节视频讲解

10.6.1 Flask Web 架构

编写程序文件 app.py，使用 Flask 框架创建一个 Web 项目，设置不同的 URL 参数对应的 HTML 模板文件。文件 app.py 的具体实现代码如下：

```python
from flask import Flask  # 导入Flask模块
from flask import render_template #导入模板函数
import _thread,time

from flask import request
from servers.data import main
from servers import xinzi,xueli,jinyan
from flask_sqlalchemy import SQLAlchemy#导入SQLAlchemy模块,连接数据库
from sqlalchemy import or_
import config#导入配置文件
app = Flask(__name__)#Flask初始化

app.jinja_env.auto_reload = True
app.config['TEMPLATES_AUTO_RELOAD'] = True
app.config.from_object(config)#初始化配置文件
db = SQLAlchemy(app)#获取配置参数,将和数据库相关的配置加载到SQLAlchemy对象中
```

```python
from models.models import Data    #导入user模块
#创建表和字段

@app.route('/')
def first():
    return render_template("input.html")

@app.route('/list')#定义路由
def list():#定义hello_world函数
    # data = Data.query.all()
    page = int(request.args.get('page', 1))
    per_page = int(request.args.get('per_page', 2))
    key = request.args.get('key', '')
    # Data.address.like("%{}%".format('java'))
    data = Data.query.filter(or_(Data.post.like("%{}%".format(key)),
                        Data.company.like("%{}%".format(key)),
                        Data.address.like("%{}%".format(key)),
                        Data.salary_max.like("{}".format(key)),
                        Data.salary_min.like("{}".format(key)) )).
paginate(page, 12, error_out=False)
    return render_template("data.html", datas=data,key=key)


@app.route('/search')
def search():
    kw =request.args.get("kw")
    city =request.args.get("city")

    # main(kw, city)
    # 创建两个线程
    try:
        _thread.start_new_thread(main, (kw, city, 1,))
        _thread.start_new_thread(main, (kw, city, 51,))
    except:
        print("Error: 无法启动线程")

    time.sleep(50)
    xz = xinzi.xinzi()
    xl = xueli.xuelifun()
    jy = jinyan.jinyanfun()
    return render_template('h.html', **locals())


#通过url传递信息
@app.route('/chart')
def charts():
    xz =xinzi.xinzi()
    xl = xueli.xuelifun()
    jy = jinyan.jinyanfun()
    return render_template('h.html',**locals())


#通过url传递信息
```

```python
@app.route('/xinzi')
def xinzitest():
    a = xinzi.xinzi()
    print(a)
    return str(a)

@app.route('/xueli')
def xuelitest():
    data = xueli.xuelifun()
    print(data)

    return str(data)

if __name__ == '__main__':
    app.run()
```

10.6.2　Web 主页

在本 Flask Web 模块中，Web 主页对应的 HTML 模板文件是 input.html，功能是提供一个表单供用户分别输入"岗位名称"和"搜索的省份"，单击"搜索"按钮后会根据用户输入的职位条件用爬虫抓取目标网站中的招聘信息。文件 input.html 的具体实现代码如下：

```
{% extends "base.html" %}
{% block content %}

    <style>
    body{
       background:
url('{{ url_for('static',filename='img/backgroun.jpg') }}');
    }
    #sousuo{
       height: 500px;
    }
    </style>
    <div id="sousuo"style="text-align: center ; padding: 5px">
          <form method="get" action="/search">
                <div class="form">
                     <p>岗位名称</p>
                     <input class="form-name"placeholder="java 软件工程师" name="kw" type="text" autofocus>
                </div><br>
                <div class="form" style="margin-top: 30px;">
                     <p>搜索的省份</p>
                     <input class="form-name"placeholder="例如:全国,江苏省" name="city" type="text" autofocus>
                </div><br>
                <input type="submit" value="搜索" class="btn" />
          </form>
    </div>

{% endblock %}
```

执行后的 Web 主页效果如图 10-2 所示。

图 10-2　Web 主页

10.6.3　数据展示页面

在本 Flask Web 模块中，数据展示页面对应的 HTML 模板文件是 data.html，功能是获取在 MySQL 数据库中保存的招聘信息，并通过表格分页的形式展示这些招聘信息。文件 data.html 的具体实现代码如下：

```
{% extends "base.html" %}
{% block content %}
    <form method="get" action="/list">
        <div class="input-group col-md-3" style="margin-top:0px; margin-left:75%; positon:relative">
            <input type="text" class="form-control" placeholder="请输入要搜索的内容" name="key"/>
            <span class="input-group-btn">
                <button class="btn btn-info btn-search" style="margin-left:3px">搜索</button>
            </span>
        </div>
    </form>


<table class="table table-bordered">
<tr>
    <th>职位</th>
    <th>公司</th>
    <th>城市</th>
    <th>最低薪资</th>
    <th>最高薪资</th>
    <th>发布日期</th>
</tr>
    {% for a in datas.items %}
    <tr>
        <td>{{ a.post }}</td>
        <td>{{ a.company }}</td>
        <td>{{ a.address }}</td>
        <td>{{ a.salary_min }}</td>
```

```
                <td>{{ a.salary_max }}</td>
                <td>{{ a.dateT }}</td>
            </tr>
    {% endfor %}
    </table>


    <div style="text-align: center">
        <nav aria-label="Page navigation example">
        <ul class="pagination justify-content-center">

            {% if datas.has_prev %}
                <li class="page-item">
            <a class="page-link" href="/list?page={{ datas.prev_num }}" aria-label="Previous">
                    <span aria-hidden="true">&laquo;</span>
                    <span class="sr-only">Previous</span>
                </a>
            </li>
            {% endif %}

            {% for i in datas.iter_pages() %}
                {% if i == None %}
                    <li class="page-item"><a class="page-link" href="#">...</a></li>
                {% else %}
                    <li class="page-item"><a class="page-link" href="/list?page={{ i }}&key={{ key }}">{{ i }}</a></li>
                {% endif %}
            {% endfor %}

            {% if datas.has_next %}
                <li class="page-item">
            <a class="page-link" href="/list?page={{ datas.next_num }}&key={{ key }}" aria-label="Next">
                    <span aria-hidden="true">&raquo;</span>
                    <span class="sr-only">Next</span>
                </a>
            </li>
            {% endif %}
        </ul>
        </nav>
        当前页数：{{ datas.page }}
        总页数：{{ datas.pages }}
        一共有{{ datas.total }}条数据
        <br>
    </div>
{% endblock %}
```

执行后的数据展示页面效果如图10-3所示。

职位	公司	城市	最低薪资	最高薪资	发布日期
Python工程师	东莞市诚誉商务信息咨询有限公司	东莞-南城区	9000	10000	05-30发布
现货操盘手（餐补+双休）	广州玖富网络科技有限公司	广州-天河区	7000	14000	05-29发布
Python开发	广州大白互联网科技有限公司	广州-海珠区	10000	18000	05-30发布
操盘手/交易员（外汇现货）	广州玖富网络科技有限公司	广州-天河区	7000	14000	05-29发布
Python开发工程师	广州回头车信息科技有限公司	广州-天河区	15000	30000	05-30发布
操盘手/交易员（天河+包吃）（职位编号：7）	广州玖富网络科技有限公司	广州-天河区	7000	14000	05-29发布
Python开发工程师	广东广宇科技发展有限公司	佛山-南海区	7000	8500	05-30发布
生物信息高级工程师	广州复能基因有限公司	广州	10000	15000	05-29发布
python教研	三七互娱	广州	8000	12000	05-30发布
生物信息工程师	广州复能基因有限公司	广州	5000	8000	英语良好
Python研发工程师	卓望数码技术（深圳）有限公司	广州-天河区	20000	25000	05-30发布

图 10-3　数据展示页面效果

10.6.4　数据可视化页面

在本 Flask Web 模块中，数据可视化页面对应的 HTML 模板文件是 h.html，功能是根据对 MySQL 数据库中保存的招聘信息的统计结果，使用 Highcharts 绘制统计图表。文件 h.html 的具体实现代码如下：

```
{% extends "base.html" %}
{% block content %}
   <style>
   body{
      background:
url('{{ url_for('static',filename='img/backgroun.jpg') }}');
   }
   </style>

   <div id="xinzi"></div>
   <table align="center">
   <tr>
      <td>
         <div id="xueli"></div>
      </td>
      <td>
         <div id="jinyan"></div>
      </td>
   </tr>
   </table>



   <script type="text/javascript">
   $(document).ready(function() {
      var chart = {
```

```javascript
        type: 'column',
        backgroundColor: 'rgba(0,0,0,0)'
    };
    var title = {
        useHTML: true,
        style: {
            color: '#000',        //字体颜色
            "fontSize": "29px",   //字体大小
            fontWeight: 'bold'
        },
        text: '工资分布图'
    };
    var subtitle = {
        text: '51job.com'
    };
    var xAxis = {
        categories: ['0~5k','5~8k','8k~11k','11k~14k','14k~17k','17k~20k','20k~23k','23K 以上'],
        crosshair: true
    };
    var yAxis = {
        min: 0,
        title: {
            text: '岗位数'
        }
    };
    var tooltip = {
        headerFormat: '<span style="font-size:10px">{point.key}</span><table>',
        pointFormat: '<tr><td style="color:{series.color};padding:0">{series.name}: </td>' +
            '<td style="padding:0"><b>{point.y:.1f} 个</b></td></tr>',
        footerFormat: '</table>',
        shared: true,
        useHTML: true
    };
    var plotOptions = {
        column: {
            pointPadding: 0.2,
            borderWidth: 0
        }
    };
    var credits = {
        enabled: false
    };

    var series= [{
        name: '岗位数',
        data: {{ xz|tojson }}
    }];

    var json = {};
    json.chart = chart;
```

```javascript
            json.title = title;
            json.subtitle = subtitle;
            json.tooltip = tooltip;
            json.xAxis = xAxis;
            json.yAxis = yAxis;
            json.series = series;
            json.plotOptions = plotOptions;
            json.credits = credits;
            $('#xinzi').highcharts(json);

    });
</script>
<script type="text/javascript">
    $(document).ready(function() {
    let chart = {
        plotBackgroundColor: null,
        plotBorderWidth: null,
        plotShadow: false,
        backgroundColor: 'rgba(0,0,0,0)'
    };
    let title = {
        useHTML: true,
         style: {
            color: '#000',         //字体颜色
            "fontSize": "29px",    //字体大小
            fontWeight: 'bold'
         },
        text: '学历占比情况'
    };
    let tooltip = {
        pointFormat: '{series.name}: <b>{point.percentage:.1f}%</b>'
    };
    let plotOptions = {
      pie: {
         allowPointSelect: true,
         cursor: 'pointer',
         dataLabels: {
            enabled: true
         },
         showInLegend: true
      }
    };
    let series= [{
       type: 'pie',
       name: '学历',
       data: {{xl|tojson}}
    }];

    let json = {};
    json.chart = chart;
    json.title = title;
    json.tooltip = tooltip;
    json.series = series;
```

```
      json.plotOptions = plotOptions;
      $('#xueli').highcharts(json);
  });
</script>
<script type="text/javascript">
  $(document).ready(function() {
    let chart = {
        plotBackgroundColor: null,
        plotBorderWidth: null,
        plotShadow: false,
        backgroundColor: 'rgba(0,0,0,0)'
    };
    let title = {
      useHTML: true,
        style: {
          color: '#000',         //字体颜色
          "fontSize": "29px",    //字体大小
          fontWeight: 'bold'
        },
      text: '工作年限要求'
    };
    let tooltip = {
      pointFormat: '{series.name}: <b>{point.percentage:.1f}%</b>'
    };
    let plotOptions = {
      pie: {
        allowPointSelect: true,
        cursor: 'pointer',
        dataLabels: {
          enabled: true
        },
        showInLegend: true
      }
    };
    let series= [{
      type: 'pie',
      name: '工作年限要求',
      data: {{jy|tojson}}
    }];

    let json = {};
    json.chart = chart;
    json.title = title;
    json.tooltip = tooltip;
    json.series = series;
    json.plotOptions = plotOptions;
    $('#jinyan').highcharts(json);
  });
</script>

{% endblock %}
```

执行后的数据可视化页面效果如图10-4所示。

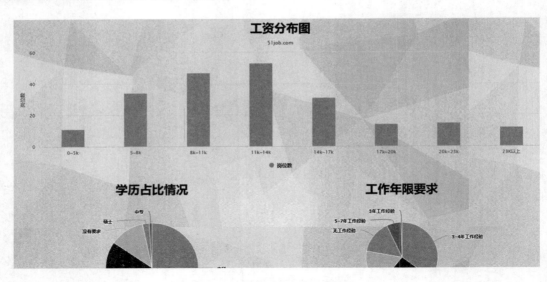

图 10-4　数据可视化页面效果